Fluidization and Fluid Particle Systems: Recent Research and Development

Hamid Arastoopour, *Volume Editor*

John C. Chen, Ye Mon Chen, Leon Glicksman, Desmond King, and Wen-Ching Yang, *Volume Co-Editors*

Mengtian Bai	H. Hatano	K. Shakourzadeh
Bernard Barry	D.M. Heyes	D.L.O. Smith
J. Baxter	B.P.B. Hoomans	Madhava Syamlal
Chia-Min Chen	Matthew R. Hyre	H. Takeuchi
Wei-Yin Chen	J.H. Kim	Stephen Tallon
N. Choueau	J.A.M. Kuipers	R.B. Thorpe
Clive E. Davies	Mooson Kwauk	K. Tsuchiya
S. Faderani	J. -F. Large	U. Tüzün
L.T. Fan	Lii-Ping Leu	Cor M. van den Bleek
Leon R. Glicksman	T. Masuyama	John van der Schaaf
P. Guigon.	S. Sakurai	W. P. M. van Swaaij
	Jaap C. Schouten	

AIChE Staff
Maura N. Mullen, Managing Editor
Armand Veneziano, Cover Design

AIChE Symposium Series

Number 318   1998   Volume 94

Published by
American Institute of Chemical Engineers

3 Park Avenue   New York, N.Y. 10016-5901

© 1998
American Institute of Chemical Engineers (AIChE)
3 Park Avenue
New York, N.Y. 10016-5901
U.S.A.

AIChE shall not be responsible for statements or opinions advanced in their papers or printed in their publications.

Library of Congress Cataloging-in-Publication Data

Fluidization and fluid particle systems : recent research and development / Hamid Arastoopour, volume editor ; John C. Chen ... [et al.], volume co-editors.
    p. cm. -- (AIChE symposium series, ISSN 0065-8812 ; no. 318, v. 94)
    "Presented in twelve sessions at the AIChE Annual Meeting in Los Angeles, California, November 16, through 21, 1997" --
Includes index.
ISBN 0-8169-0774-9
    1. Fluidization--Congresses. 1. Arastoopour, Hamid. II. Chen John C., 1934- .   III. American Institute of Chemical Engineers. Meeting (1997 : Los Angeles, Calif.)   IV. Series: AIChE symposium series ; no. 318.
TP156.F65F5736   1998
660'.284292--dc21   98-36294
                                                                         CIP

    All rights reserved whether the whole or part of the material is concerned; specifically those of translation, reprinting, re-use of illustrations, broadcasting, electronic networks, reproduction by photocopying machine or similar means, and storage of data in banks.
    Authorization to photocopy items for internal use, or the internal or personal use of specific clients, is granted by AIChE for libraries and other users registered with the Copyright Clearance Center Inc., provided that the $3.00 base fee + $1.50 per page is paid directly to CCC, 222 Rosewood Drive, Danvers, MA 01923. This consent does not extend to copying for general distribution, for advertising, or promotional purposes, for inclusion in a publication, or for resale.
    Articles published before 1978 are subject to the same copyright conditions and the fee is $3.50 for each article. AIChE Symposium Series fee code: 0065-8812/1998.

FOREWORD

This volume of the AIChE Symposium Series continues the tradition of an annual volume presenting recent research and development in fluidization and fluid-particle systems. The papers were selected by peer reviews from those presented in twelve sessions at the AIChE Annual Meeting in Los Angeles, California, November 16 through 21, 1997.

This year's symposium volume contains papers that cover a wide range of topics related to fluid-particle systems and fluidization including:

- Fundamentals of Fluidization and Fluid-Particle Systems
- Particle Interaction
- Circulating Fluidized Bed
- Modeling and Computer Simulation of Fluid-Particle Systems
- Advances in Fluid-Particle Flow Measurements, Handling and Processing

The first article is a fluidization plenary paper entitled "Exploring the Multi-Phase Nature of Fluidization" by Professor Mooson Kwauk of the Institute of Chemical Metallurgy, Academia Sinica, China, who received the 1997 Fluidization Lectureship Award sponsored by Fluor Daniel Foundation.

I would like to acknowledge all the chairs and co-chairs of the sessions for their careful selection of papers for the Los Angeles meeting. I would also like to thank my co-editors, AIChE Managing Editor Maura N. Mullen, Dr. Amir Riahi, and the fluidization particle technology research team at Illinois Institute of Technology, whose critical review and further evaluation of the papers ensured the high quality of the papers in this volume. Finally, I would like to express my thanks to my assistant, Simone Bradford, without whose diligence, quality control and hard work, this volume would not have been possible.

Hamid Arastoopour, *Volume Editor*
Professor and Chairman
Department of Chemical and Environmental
Engineering
Illinois Institute of Technology
10 W. 33rd Street
Chicago, Illinois 60616

CONTENTS

Foreword iii

PLENARY PAPER

Exploring the Multi-Phase Nature of Fluidization
Mooson Kwauk..1

RESEARCH AND DEVELOPMENT PAPERS

The Influence of a Particle Size Distribution on the Granular Dynamics of Dense Gas-Fluidized Beds: A Computer Simulation Study
B.P.B Hoomans, J.A.M. Kuipers, and W.P.M van Swaaij 15

Determination of Lateral Dispersion Coefficients in the Dilute Region of Fluidized Beds
Matthew R. Hyre and Leon R. Glicksman 20

Long-Range Connectivity in Slow-Shearing Granular Flows
J. Baxter, U. Tüzün, and D.M. Heyes 25

Motion of Individual FCC Particles and Swarms in a Circulating Fluidized Bed Riser Analyzed via High-Speed Imaging
H. Hatano, H. Takeuchi, S. Sakurai, T. Masuyama, and K. Tsuchiya...........................31

Mechanism of Solid Flow in a Closed Loop Circulating Fluidized Bed with Secondary Air Injection
J.H. Kim, K. Shakourzadeh, and J.F. Large 37

Non-Intrusive Solids Velocity Measurement in a Downcomer
Stephen Tallon, Clive E. Davies, and Bernard Barry 42

Experimental Observation of Pressure Waves in Gas-Solids Fluidized Beds
John van der Schaaf, Jaap C. Schouten, and Cor M. van den Bleek 48

Higher Order Discretization Methods for the Numerical Simulation of Fluidized Beds
Madhava Syamlal 53

A Drift-Flux Model for Flow of Nearly-Buoyant Coarse Granular Solids in Liquids
S. Faderani, U. Tüzün, D.L.O. Smith, and R.B. Thorpe 58

Modeling Bed-Load Transport by the Master-Equation Approach
L.T. Fan, Mengtian Bai, and Wei-Yin Chen 63

Mass Transfer and Flow Regimes in Three-Phase Magnetic Fluidized Beds
Chia-Min Chen and Lii-Ping Leu 70

Characterization of Attrition Properties by a Shear Test
N. Chouteau, P. Guigon, and J-F Large ....…..……….75

Index 80

Plenary Paper

Exploring the Multi-Phase Nature of Fluidization

Mooson Kwauk
Institute of Chemical Metallurgy, Academia Sinica, Beijing 100080, China

When Fluidization was first studied in the 1940's, it was treated as a global phenomenon. People were concerned with primary variables, such as, the velocity at which particles start to fluidize and how much the solids bed expand with velocity, etc. When chemical reaction took place in a fluidized bed, it was treated in the manner of its kin, the fixed-bed reactor, in terms of yield, conversion, and other factors which obtain at the inlet and outlet of the fluidized bed. With intensification of application, however, we require more precise knowledge about the in situ heterogeneity of the fluidized systems and we need to know what actually happens to individual particles as they react with the gas. This presentation explores the following aspects of the multi-phase nature of fluidization:
   Morphological Changes of Reacting Particles
   Heterogeneity/Homogeneity in Particle-Fluid Two-Phase Flow
   Improvement of G/L Conatacting with Fluidized Solids
   Fluidization of Particles with Appreciable Inter-Particle Forces

When fluidization was first studied in the 1940's, it was treated as a global phenomenon, and people were concerned with such primary variables as the velocity at which particles start to fluidize, how much the solids bed expands with velocity, etc. When chemical reaction took place in a fluidized bed, it was once treated much in the manner of its kin, the fixed-bed reactor, in terms of yield, conversion, and other factors which obtain at the inlet and outlet of the fluidized bed. With intensification of application, however, we need more precise knowledge of the *in situ* heterogeneity of the fluidized systems and we need to know what actually happens to *individual* particles as they react with gas. I feel honored to be invited to share with you some of our in-house studies, which explore a few aspects of such multi-phase nature of fluidization.

MORPHOLOGICAL CHANGES OF REACTING PARTICLES (Z. Cao, M. Shao, X. Ma)

In the case of a reacting solid particle, it is often geometrically simplified to a sphere or a spherical agglomerate of smaller spheres. While the supposition of a sphere simplifies mathematical modeling, it fails to reflect what actually happens to the particle during processing. Two attempts were made to reveal the morphological changes of a reacting solid particle: observation of a reacting solid particle while it is acoustically levitated, and observation of the solid surface in a microreactor under a scanning electron microscope.

When a single particle is levitated in a flowing reacting medium, the whole reaction process of a given particle could be traced visually while eliminating interactions from other particles as well as from the reactor walls. Acoustic levitation is used in such containment of single particles in fluid media. To counteract the dependence of sound wave propagation on the changing properties of the fluid medium during the course of reaction, a resonance tracking function is required to compensate the effects due to medium changes in order to maintain stable levitation of the particles. Theoretical analysis showed that there are two factors which can be manipulated to maintain resonance in the chamber: the length L of the chamber and the frequency f of the sound wave.

Figure 1-1 shows the single-axis acoustic levitation reactor together with its resonance tracking system. A piezoceramic transducer operated at 20 kHz is used as the sound source. Phase shift between the input and output of the chamber is used as the measurement multitude of minute nuclei of nascent oxide. Above this thin layer is a second, more or less compact layer consisting of agglomerated grains due to protracted residence, and above this intermediate layer is another layer of discrete younger grains, rectangular or hexagonal, stick-like or whiskers, according to reaction conditions, while Fe and O diffuse countercurrently to each other.

Mathematically, attempts were made to adapt existing methodology to modeling complex particle morphology, e.g., simplification of the Fourier analysis from cosine series, which employs both amplitude coefficients and phase angles, to pure sine series (**Figure 1-9**) which avoids using phase angles, and fractal theory. But as shown in the above physical model, much ground needs yet to be covered between mathematical techniques and the results derived from the levitation reactor and the ESEM microreactor.

HETEROGENEITY/HOMOGENEITY IN PARTICLE- FLUID TWO-PHASE FLOW
(J. Li, D. Liu, G. Xu)

In 1946, when I was carrying our my dissertation research in fluidization, I was greatly fascinated by the different behaviors of particles while fluidized by a liquid and by a gas. Liquid/solid fluidization is quite homogeneous, and gas/solid fluidization is characterized by bubbling and slugging. The former was then designated "particulate" and the latter, "aggregative." Such distinction between particulate and aggregative fluidization has since commanded much attention, though relatively little has been discussed as to their common origin, or to their transition -- abrupt, gradual, or in what manner.

In later years, while developing new processes treating vast tonnages of low-grade ores, revelation from these two types of fluidization led me to realize that the conventional bubbling gas/solid fluidized bed consumes too much energy for the circulating gas and yet provides inadequate contact between gas and solids because of gas bypassing through bubbling. The solid particles need to be *dispersed discretely* as in particulate fluidization for liquid/solid systems. The answer was dilute-phase operation, which I adopted in most of the fluidized-bed metallurgical processes that I devised. A more basic question, however, was to look at fluidization of solid particles with a fluid having properties intermediate between those of a gas and of a liquid. I thought of the use of supercritical fluids, and in 1957 I collated data on fluids spanning the common liquids and gases, including gases under pressure.

In 1984, Jinghai Li came to me to study fast fluidization. I asked him to analyze the problem by resolving the particle-fluid system into three scales of interaction: a **micro** scale of the dimension of the individual particles, a **meso** scale of the dimension of the strands (or clusters) and a **macro** scale of the dimension of the equipment (**Figure 2-1**). A set of equations was written, but the constraints seemed to be one less than the variables specified. Jinghai believed that the phase structure of a particle-fluid system is an expression of its stability, which is determined by certain minimal energy of the system. He tried quite a number of formulations of the energy term, until he finally pinned down the *minimal energy for suspension and transport of the solids*.

Convergence in computation then posed another problem. When the computer began to yield smooth curves, we discovered that at some singular point the computed variables were discontinuous. This was soon understood to correspond to the common phenomenon of "choking." For solids with good fluidizing behavior, such as the Geldart Group A powders, choking is delayed to higher gas velocities. For normal liquid/solid systems, however, choking was found to be absent. Finally, we computed fluidization with supercritical CO_2, and found transitional behavior which varies continuously from G/S to L/S systems as pressure increases to near the critical point. We called the above model, the *Energy-Minimization Multi-Scale* model, or for short, the *EMMS* model.

Figure 2-2 compares the computed results for the FCC/air system against those for the glass/water system, to illustrate the disparate behaviors of G/S and L/S fluidization. The first two insets on the left-hand side show the change of voidage ε_f, ε_c and ε (dilute-, dense-phase and average, respectively) and cluster-phase fraction f for the FCC/air system. In the corresponding insets for the L/S system to the right, however, the three voidages ε_f, ε_c and ε are identical, and the cluster-phase fraction f is zero, indicating the absence of clusters throughout the velocity range of U_g, that is, fluidization is homogeneous.

variable, and under the control of this feed back loop, the chamber length L is maintained at an optimum resonant value and particles are levitated stably during the course of medium changes. **Figure 1-2** shows the heating, temperature measurement and image sampling system. The heating source is a CO_2 laser. The particle is irradiated and its temperature is measured by an infrared thermometer. Based on the error of the temperature signal, a microcomputer-based digital PID controller generates a control signal to vary the laser power required to maintain the particle at the desired temperature.

Chalk, plastics, glass and aluminum particles in the size range of 3 to 5 mm, as well as a 4-mm steel sphere have been successfully levitated in the apparatus. The ability of radial and vertical focusing was proved by the measured sound pressure distribution in the resonance chamber, as shown in **Figure 1-3**, for axial sound pressure distribution in the upper part of the chamber at ambient temperature with a particle levitated at the lowest sound pressure node. The distance between the adjacent pressure nodes was found approximately to be half of the calculated wavelength $\lambda_{0,1}$ for a $(0,1)$ normal mode standing wave, which is the desired mode for providing radial focusing in the chamber.

For a levitated droplet of NH_4Cl solution, nascent crystals could be seen growing inside the droplet while crust formation proceeds simultaneously. Also evident are whisker/dendrite growths outside the crust extending into free space.

On an even finer scale, for what the human eye cannot discern and differentiate, the surface texture of the particle is often left unaccounted for. However, different individual microstructure of the same materials may induce diverse behaviors for the reacting particle. Due to its outstanding technological importance and extensive prior studies, the oxidation of iron was chosen as a model system for illustrating the significance of morphological changes at the surface. We shall describe experiments using *environmental scanning electron microscopy* (ESEM), in which specimens could be examined directly and continuously in a controlled reacting gas environment.

The KYKY 1500 ESEM, which we built in Beijing as shown in **Figure 1-4**, features a three-step vacuum system, a newly designed detector and a hot stage (the microreactor), with its gas supply system, graphics system and external temperature control system, all tailored specifically for dynamic and *in situ* observation and video recording in the microreactor. The specimens used in the present work were cylinders of 99.98 wt% electroformed iron, 2.4 mm in diameter, with polished surface. A VIDAS-25 graphics system, linked to the ESEM, sequentially recorded and stored the experimental data. Some specimens oxidized in the ESEM were further examined by using post-oxidation techniques, such as FESEM (AMRAY 1910 FE), petrographic analysis and Auger energy spectroscopy (AES PHI-610), for identifying and characterizing the oxidation products.

Three sets of experimental conditions were chosen to observe the different formation processes for different oxide grains of iron. In the **first** set of conditions, carried out at 500° C, argon was first introduced into the specimen chamber at about 200 Pa, until the desired temperature was reached and stabilized. Then air was mixed into the argon stream to an oxygen partial pressure $P_O=3.2$ Pa. **Figure 1-5** shows the successive changes at the surface, until finally new nuclei became whiskers. The **second** set of experiments was performed at 700°C at $P_O=44$ Pa. **Figure 1-6** shows that successive layers of particulate oxide grains formed from the metal substrate, which finally became whiskers, before eventual agglomeration. The **third** set of experiments was carried out at 600°C with programmed gas entry. The specimen was first heated with air with $P_O=200$ Pa, to form an initial oxide layer. When the desired temperature had been reached, argon was supplied to substitute air, with a residual $P_O=8.5 \times 10^{-4}$ Pa. The micrographs in **Figure 1-7** show a different process for iron oxide growth, resulting in relatively large rectangular grains which slowly changed to hexagonal grains.

The above three sets of experiments show that when oxidation was carried out under similar oxygen partial pressures but at different temperatures, the iron oxide products are totally different: thin, flat, leaf-like whiskers at 500°C (Figure 3(f)); agglomerated whiskers at 700°C (Figure 4(f)); rectangular and hexagonal grains at 600°C (Figure 7(f)). These oxides of different morphologies imply different growth mechanisms and reaction kinetics, thus calling for different modeling techniques. **Figure 1-8** presents a plausible physical model for the oxidation of iron on the basis of the ESEM observations described above. Next to the Fe surface is a thin layer of nucleation consisting of a

Figure 2-3 shows the gradual transition of the homogeneous glass/water fluidization to the highly heterogeneous, or aggregative glass/air fluidization, as the particle/fluid density ratio ρ_p/ρ_f increases from water through ethyl ether, and carbon dioxide under different stages of decreasing pressure down from its critical condition, to atmospheric air. The appearance and gradual growth of the two-phase structure is evident in the order of the fluids listed.

While Figure 2-2 demonstrates from our modeling the *disparate* nature between G/S and L/S fluidization, Figure 2-3 shows *continuity* in particle-fluid behavior through properly selected intermediate fluids, thus reconciling through modeling the phenomenological discrimination between particulate and aggregative fluidization.

Such computation needs to be corroborated by experiments. In 1990, Dejin Liu, was given the task of experimental investigation of fluidizing solid particles with supercritical CO_2. His experiments were conducted in a stainless steel column, 26 mm i.d. by 2 m high, connected to a CO_2 circulating, heating and regulating system as shown in **Figure 2-4**. These were supplemented with additional experiments using liquids less viscous and much more viscous than water, in another L/S system with a column of similar dimensions. Voidage of the fluidized solid particles and their fluctuations were measured by means of special optical fiber probes located along the column height, with on-line data acquisition and processing. Nine kinds of solid particles were used, ranging in **density** from the lightest for silica gel to steel, and in **size** from the smallest for alumina to steel balls. Seventeen species of liquid were used, ranging in **density** from the lightest for CO_2 at ambient conditions through CO_2 at 9.4 MPa, n-hexane, water to 70% aqueous glycerol, and in **viscosity** from the least viscous for CO_2 at ambient conditions, through CO_2 at 9.4 MPa, n-hexane, water to 70% aqueous glycerol.

While the EMMS model establishes on broad principles the distinction as well as reconciliation between particulate and aggregative fluidization, Dejin's experiments scan the more **specific** behaviors of the fluidized systems in order to supply other more descriptive criteria. These behaviors fall into the following three categories.

Spatio-temporal Voidage Fluctuation

Figure 2-5 shows typical *fluctuating voidages* for fluidizing particles with CO_2 under pressure: sand with CO_2 at 0.1 MPa, ion exchange resin at 3 MPa, and FCC catalyst at 6 MPa, in the order of increasing homogeneity. The simplest statistical parameter, the departure of the local and instantaneous voidage ε_i from its average value ε is used, in the form of the standard deviation:

$$\sigma = \sqrt{\frac{1}{n}\sum_{n=1}^{n}(\varepsilon_i - \varepsilon)^2}$$

The most heterogeneous fluidization occurs in slugging, for which voidage alternates between a dilute phase with voidage approaching unity and a dense phase with voidage approaching that at minimum fluidization ε_{mf}, with their respective fractional occurrences of f and $(1-f)$. According to Dejin's analysis, the maximum value of $\sigma_{slugging}$ occurs at $f = 1/2$, from which $\sigma_{max} = \frac{1-\varepsilon_{mf}}{2}$ and $\varepsilon_{\sigma_{max}} = \frac{1+\varepsilon_{mf}}{2}$.

A relative measure, called *heterogeneity index* δ, can thus be defined:

$$\delta = \frac{\sigma}{\sigma_{max}} = \frac{2\sigma}{1-\varepsilon_{mf}}.$$

The range of variation of δ is from 0 to 1:

$$0 < \delta < 1$$

max homogeneity   max heterogeneity

Expansion of Fluidized Bed with Velocity

For the velocity range between U_{mf} and U_t, *bed expansion* can be described in a global sense by the area under the voidage-*versus*-velocity curve, as shown in **Figure 2-6**:

$$A_R = \int_{U_{mf}}^{U_t} \varepsilon \, dU$$

where, for particulate fluidization, $U = U_t \varepsilon^n$, substitution and integration give

$$A_P = \frac{n}{n+1}U_t\left[1-\left(\varepsilon_{mf}\right)^{\frac{n+1}{n}}\right]$$

As fluidization becomes less homogeneous, particles aggregate and the bed expands less, resulting in a smaller area, $A_R < A_P$. This deviation of A_R from A_P can be taken as a measure of the *global nonideality* f_h of the fluid-particle system:

$$f_h = \frac{A_P - A_R}{A_P} = 1 - \frac{\int_{U_{mf}}^{U_t}\varepsilon\,dU}{\frac{n}{n+1}U_t\left[1-\left(\varepsilon_{mf}\right)^{\frac{n+1}{n}}\right]}$$

The range for the variation of f_h is also from 0 to 1:

$$0 < f_h < 1$$
max homogeneity   max heterogeneity

Figure 2-7 shows three sets of experimental results for the variation of *bed expansion* and of *local heterogeneity* with fluid velocity for three kinds of particles. The lower set is for silica gel, which is relatively light, showing that the bed expands almost linearly with fluid velocity for n-hexane and CO_2 at 9, 8 and 6 MPa, while the local heterogeneity index δ starts from a very low value of 0.025, increasing in the order of the fluids mentioned. In this set, with CO_2 under progressively decreasing pressures, from 4 MPa to 0.1 MPa, bed expansion becomes more and more difficult as particles aggregation becomes more pronounced, while the local heterogeneity increases in that order up to a high value of 0.7. The middle set is for ion exchange resin, which is denser, showing that linear bed expansion is even poorer, and that the local heterogeneity index δ starts from a much higher value of 0.07. The upper set is for the heaviest particles tested, steel balls, showing that only aqueous glycerol solutions can maintain linear bed expansion, and even with CO_2 at 5 MPa, the local heterogeneity index is as high as 0.9.

Role of Particle and Fluid Properties

The choice of the dimensionless numbers involving *particle and fluid properties* to demarcate particulate from aggregative fluidization depends greatly on how well they fit experimental data. Dejin proposed what he called the *discrimination number*:

$$Dn = \left(\frac{Ar}{Re_{mf}}\right)\left(\frac{\rho_p - \rho_f}{\rho_f}\right)$$

Figure 2-8 shows the interrelationship between δ, f_h and Dn. Though they are well correlated, they belong, nevertheless, to totally different behaviors:

- δ *local heterogeneity* for spatio-temporal *fluctuation of voidage*
- f_h *global nonideality* for suppressed *bed expansion*
- Dn *discriminating* the contributing influences of *particle and fluid properties*

These criteria, therefore, represent but different quantified aspects of the same phenomena, each giving its own quantitative measure as to how far a particle-fluid system is removed from aggregative fluidization and how close it is to particulate fluidization.

As originally formulated, the EMMS model was in the form of an optimization problem. Solution depended on the use of the GRG2 (General Reduced Gradient) program, which was time consuming and often did not converge easily. To facilitate the application of the EMMS model to engineering problems, attempts were made for simpler solutions. Guangwen Xu (1997) devised an analytical approach which reduces the computation time from *minutes* and *hours* to *seconds*.

Fifty years ago, as a student, I studied fluidization and observed the liquid/solid pairs that fluidize particulately and the gas/solid pairs that fluidize aggregatively. Not only out of curiosity to pursue the unknown, but also driven by practical motives for developing new processes and improving existing processes, I have been, for the better part of my life, looking for the link between the two. The answer has come through the efforts of three generations of investigators ---- my teacher, myself and my two students.

Through such a knowledge of the common basis of particulate *vis-a-vis* aggregative fluidization, we may well be able to better fluidize a solid, be that with a liquid or with a gas. Could we, for instance, possibly "particulatize" a gas/solid system in the direction of a liquid/solid system, for instance, in devising a feed/transport/injection system for powdered coal, which operates as smoothly as coal-water-mixture yet without the loss of gasification or combustion efficiency due to the presence of the carrying fluid, water. Or could we think of some way of increasing the throughput of liquid/solid fluidization, e.g., fluidized leaching and/or washing, by the incorporation of small quantities of ferromagnetic micro particles for operation in some kind of a magnetic environment, strategically emplaced in the processing vessel, with alternating or even electrically moving field. We may have exhausted our present arsenal of concepts, and need to create new weaponry of ideas.

IMPROVEMENT OF G/L CONTACTING WITH FLUIDIZED SOLIDS
(X. Ma, F. Ouyang, Y. Wu, L. Cheng)

Magnetofluidization was first used in the containment of iron-based catalyst particles in G/S fluid-bed reactors, for instance, in ammonia synthesis and water-gas shift reaction. The magnetic filed is capable of holding back the ferromagnetic catalyst particles against an upflowing gas stream. Applied to G/L/S systems, magnetofluidization can disintegrate gas bubbles. Whereas bubbles in G/L/S systems are quite different from bubbles for G/S systems, they share the common feature that the parallel lines of a uniform magnetic field are flexed by the presence of a void, in which there are no ferromagnetic particles, and thus create a magnetic tension oriented toward the center of the bubble, forcing the ferromagnetic particles to penetrate and fall into the bubble.

Experiments for G/L/S magnetofluidization were carried out in a two-dimensional bed as shown in **Figure 3-1**, 150 x 10 x 1000 in inside dimension. Iron granules, 0.09 to 0.37 mm in diameter, were fluidized with water doped with a common detergent to alter its surface tension. Air was injected via a solenoid valve through a 3-mm nozzle closely above the liquid distributor, to generate single bubbles. A horse-shoe electromagnet supplied a horizontal magnetic field through the frontal surface of the 2-D bed. **Figure 3-2** shows the bubble detector, which consists of a lower horizontal row of 29 photocells deployed equidistantly at 5-mm intervals, and a single horizontally movable upper probe, located 18 mm above the lower row. Illumination from the opposite side of the 2-D bed sent parallel rays through the bed front towards the bubble detector. Signals from the 30 photocells were sent to a computer for on-line processing. **Figure 3-3** shows a computer printout of the bubble pattern. **Figure 3-4** shows a series of pictures of bubble behavior under different magnetic field strengths. Without any magnetic field, H = 0 Oe, Figure 3-4a shows a large single bubble, about 9 cm in diameter, from the injected air pulse. With increasing magnetic field strength, $0 < H < 80$ Oe, Figures 3-4b and 3-4c show the big bubble breaks up into progressively smaller bubbles. Here, particles can be seen showering from the roof of the bubble to cleave it into smaller fragments.

The bubble roof is subject to both the magnetic force of particle penetration and the gravity force of the particles from above, and therefore becomes the weak point of fracture. Through formulation of the magnetic, interfacial-tension, pressure, gravitational and drag forces, an expression has been derived for computing the maximum stable bubble size. **Figure 3-5** shows the effect of magnetic field strength on *bubble diameter* with changing liquid and gas flow rates, particle diameter and gas-liquid interfacial tension. The points in this figure are experimental and the curves are computed. Similarly, **Figure 3-6** shows the effect of magnetic field strength on *bubble velocity* with the same set of changing parameters.

Figure 3-7 shows a magnetofluidized G/L/S ethanol bioreactor, which employs a magnetic distributor-downcomer consisting of an iron mesh surrounded by a collar-type magnetizing coil. The solid particles are immobilized yeast on alginate beads with molded-in magnetite powder. These ferromagnetic particles, when placed in the magnetic field, prevent their being carried upward as a scum by the CO_2 bubbles derived from fermentation which adhere to the particles. When the coil is magnetized, a nonuniform magnetic field is created above and below the distributor, resulting in the formation of all the three regimes of magnetofluidized L/S systems: particulate, chain and magnetically condensed, ordered as above according to their proximity to the magnetic distributor-downcomer.

Magnetofluidization is fairly well provided with basics studies and can be considered a propitious area for new innovative techniques.

FLUIDIZATION OF PARTICLES WITH APPRECIABLE INTER-PARTICLE FORCES
(Z. Wang, H. Li)

In the offing there will be more fluidization techniques dealing with or making use of inter-particle forces. These range from ash-agglomerating coal gasification, the preparation and processing of ultrafine particles, to granulation of powders, including the preparation of detergents coated with proteinase and the making of Chinese micro-pill herb medicines which are easier to swallow than mega-sized capsules. Such techniques constitute a class in its own, involving somewhat common mechanisms and calling for somewhat generic tools for analysis and design.

Take as an example, the study of fine particles. Fine particles are understood to be particles fine enough to show appreciable cohesion between particles and adhesion between particles and any surface in contact. The behavior of fine particles when fluidized is highly unpredictable and therefore experimental results would be more *representative* than *reproducible*. To insure a higher degree of representativeness, different types of experiments are performed on samples to look for mutual congruence: bed expansion upon fluidization, bed collapsing test, measurement of cohesiveness among particles, dynamic video recording of fluidization behavior followed by image analysis, measurement of porosity of bulk solids. The results of these individual tests are collated to sharpen the overall representativeness of the fine particle behaviors.

Figure 4-1 shows the experimental fluidized bed which measures 33 mm in inside diameter and 1000 mm in height. Bed collapsing test was performed by instantly cutting off the fluidizing air by a solenoid valve, following the descending bed surface by a Sony handycam camera and processing the collapsing curves with a PC computer. Fine particles used were goethite, titanium dioxide, talc, alumina, nickel, magnetite, zeolite, aerogel, etc., ranging in size from 0.01 to 18.1 μm and in density from 100 to 8600 kg/m^3.

Figure 4-2 shows that goethite particles, for instance, initially form a plug with a high pressure drop of 32 mm H$_2$O, much greater than the apparent weight of the particles. The plug splits and the bed channels at U=0.02 m/s, whereby the pressure suddenly drops to a low value of 3.5 mm H$_2$O. When the gas velocity is increased to 0.04 m/s, the pressure drop rises abruptly to 30 mm H$_2$O, and the bed disrupts itself and the particles are fluidized, and thereafter the bed expands with further increase in gas velocity from 180 to 345 mm, while the pressure drop remains essentially the same.

As a whole, the process of fluidizing fine particles usually involves plugging, channeling, disrupting and agglomerating, and their combinations, each highly specific and characteristic of the particles being dealt with. Beds of fine particles could generally be disrupted suddenly at some characteristic gas velocity, called the disrupting velocity, $U_{disrupt}$.

Bed collapsing behavior of fine particles generally falls into two classes. In the first case, for particles with very small bed expansion and serious channeling, their beds hardly collapse. The second class refers to particles with some appreciable expansion ratio, demonstrating the three-stage collapsing characteristics of Geldart group A particles, such as for goethite as shown in Figure 4-2. For these particles, different sizes of agglomerates of fine particles are formed, and often the hindered sedimentation stage is long.

Experiments showed that when fine particles agglomerate during fluidization, the larger agglomerates often gravitate to the bottom sometimes forming a fixed bed, the medium agglomerates are fluidized in the middle, and there is a dilute-phase at the top consisting of small agglomerates and discrete, unassociated particles. The agglomerates were photographed at different bed heights and the agglomerate sizes were measured to give the agglomerate size distribution shown in **Figure 4-3**.

The height of the agglomerate layer first increases rapidly, and then, as larger agglomerates are broken into smaller ones and elutriated out, the agglomerate layer decreases. When the elutriated agglomerates are collected and recycled repeatedly into the bed, the average size of the agglomerates would decrease and become more or less constant. These final agglomerates would fluidize homogeneously. This phenomenon suggests the use of circulating fluidized bed to achieve better fluidization of fine particles through this process of *equilibrium agglomeration*.

Fine particles tend to form globular agglomerates when kept in a heap, stored in a vessel, or while being transferred, that is, whenever relative motion exists between particles. These agglomerates, called *natural agglomerates*, are generally light and friable in structure, and possess relative close size ranges. When fluidized, the natural agglomerates undergo reorganization, or are fragmented into smaller agglomerates or even discrete particles, which reform afresh new agglomerates, the *fluidized agglomerates*.

Analysis of experimental results showed that

- The stability of natural agglomerates increases with decrease of particle size.
- Better fluidization results for multi-sized agglomerates.
- Low-density agglomerates fluidize better.
- The larger agglomerates at the bottom of a fluidized bed are usually composed of smaller parent individual particles, while the smaller agglomerates further up in the bed consist of larger members of the parent discrete particles.
- Channeling and slugging always take place for fine particles, and channels can form either directly from the discrete fine particles or from their agglomerates.

The fluidizing behavior of fine particle agglomerates may be compared to that of normal discrete particles as shown in **Figure 4-4**. The top line shows the normal fluidization of discrete noncohesive particles, spanning the velocity range from minimum fluidization U_{mf} to particle transport U_t. The bottom line shows the worst case for unfluidizable fine particles which channel or rat-hole for the entire velocity range from computed particle minimum fluidization at U_{mf} to transport of the unfluidized particles as a single agglomerate at $U_{transport}$. Then, from the upper left corner of the rectangle a line is drawn to the lower right corner to represent the locus of the disruptive velocity $U_{disrupt}$. For any sample of particles which channel at low velocity and fluidize with agglomerate formation at higher velocity, there exists a horizontal line starting from some point C, between the top and bottom lines, to intersect this $U_{disrupt}$ locus at a point equal to its disruptive velocity, to demarcate the channelling/rat-holing behavior toward the left from the agglomerate fluidizing behavior toward the right. The lower is this C-line, the greater is the fractional velocity range for channelling/rat-holing. Thus, if we graduate the ordinate between zero and unity to represent an arbitrary scale of relative cohesiveness, RC, then the area of the lefthand-side triangle above the C-line will represent the departure of the fine particles chosen from discrete particles, and the full range for RC will represent all the intermediate behaviors between zero for normal fluidization of non-cohesive particles and unity for unfluidizable fine particles:

$$0 < (RC)^2 < 1$$

normal fluidization -- channelling/rat-holing
unfluidizable non-cohesive particles
(followed by fine particles agglomerate fluidization)

Of course, the $U_{disrupt}$-locus should be experimentally determined, rather than expressed as a straight line as shown, and the departure from normal fluidization should be evaluated by integration along the U-axis.

The multi-phase nature of fluidization has provided us with a rich spectrum of challenging problems. Heterogeneity/homogeneity has been an outstanding problem for decades. We have proposed a solution, though far from being perfect. The morphology and texture studies are aimed at better modeling of solid-state changes. They call for some knowledge of non-chemical-engineering areas, viz., acoustics and microscopy. Improvement of G/L contacting with magnetofluidization is focused on the utilization of readily available basic studies. Current interest in ultrafine particle fluidization is being foreseen as an extending field to allied processes dealing with or making use of inter-particle forces. It is hard to generalize as to how to identify problems related to the multi-phase nature of fluidization. For one thing, they are usually situated at the intersections between technologies, sciences and even professions. Neither it is easy to generalize on the solution of these problems. Often they call for some knowledge of neighboring science or technology. As a whole, the multi-phase nature of fluidization calls upon the chemical engineer to know more about other technologies and to be prepared to be apprentice of other disciplines in science and technology.

The past half a century or so has seen tremendous growths and accomplishments in fluidization. But we can further benefit ourselves from fresh fertile areas of endeavors. Isolated, every new concept or innovative device/process is a step-change of breakthrough, but, taken as a whole, progress stands for continual integration of the inheritance of the accepted and imagination of the unexpected. I am sure there will be many such step-change breakthroughs in this symposium, and I sincerely wish you success in this meeting.

PRINCIPAL REFERENCES

1. Cao, Z., Liu, S., Li, Z., and Gong, M., Development of an Acoustic Levitation Reactor, *Powd. Technol.*, 69, 125-131 (1992)
2. Cao, Z., Liu, S., and Li, Z., Acoustic Levitation Reactor, *14th Intern. Congr. Acoustics*, Beijing, 1992
3. Cao, Z., Liu, S., Li, Z., Diao, Y., Pan, D., and Luo, B., The Design and Control of an Acoustic Levitation Reactor, *Chem. Reaction Eng. & Technol.*, 11(2), 120-127 (1995)
4. Cao, Z., Li, Z., and Liu, S., Resonance Tracking System for Acoustic Levitation Apparatus, Chinese Pat. Appl. CN 1056176 91103537.0 (Nov. 13, 1991)
5. Kwauk, M., Ma, X., Ouyang, F., Wu, Y., Weng, D., and Cheng, L., Magnetofluidized G/L/S Systems, *Chem. Eng. Sci.*, 47(13/14), 3467-3474 (1992)
6. Li, J., and Kwauk, M., "Particle-Fluid Two-Phase Flow – the Energy-Minimization Multi-Scale Method," Metallurgical Industry Press, China, 204 pp (1994)
7. Liu, D., Kwauk, M., and Li, H., Aggregative and Particulate Fluidization – the Two Extremes of a Continuous Spectrum, *Chem. Eng. Sci.*, 51(17), 4045-4063 (1996)
8. Ma, X., Pure Sine Descriptors for Particle Shape Analysis, Particuology '88 (*Proc. Trilat. Symp. Particuology*, Sept. 5-9, 1988), p. 7-12
9. Ma, X., and Kwauk, M., Particulate Fluidization of Ferromagnetic Particles in Uniform Magnetic Field, *CREEM Seminar* (Chem. Reaction Eng. In Extractive Metall.), Beijing, May 1985, p. 248-267; *Pacific Reg. Meet. of Fine Particle Soc. (U.S.A.)*, Honolulu, Hawaii, Aug. 1-5, 1983, Proc., ed. T. Ariman, p. 117
10. Ouyang, F., Wu, Y., Guo C., and Kwauk, M., Fluidization under External Forces (1) Magnetized Fluidization, *J. Chem. Ind. Eng. (China)*, 5(2), 206-222 (1990)
11. Shao, M., Li, H., and Kwauk, M., Morphological Changes of Oxide Grains during Oxidation of Pure Iron in an Environmental Scanning Electron Microscope, *Particle & Particle Characterization*, 14(1), 35-40 (1997)
12. Wang, Z., Kwauk, M., and Li, H., Fluidization of Fine Particles, *Chem.Eng. Sci.*, in press
13. Wilhelm, R. H., and Kwauk, M., Fluidization of Solids Particles, *Chem. Eng. Prog.*, 44, 201 (1948)
14. Xu, G., and Li, J., Analytical Solution of the Energy-Minimization Multi-Scale Model for Particle-Fluid Two-Phase Flow, *Chem. Eng. Sci.*, in press

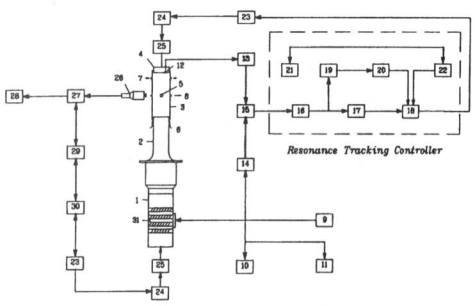

Figure 1-1. Acoustic levitation reactor with its resonance tracking and particle position control systems

1. Ultrasonic transducer, 2. Stepped horn, 3. Resonance chamber, 4. Reflector, 5. Levitated particle, 6. Gas inlet, 7. Gas outlet, 8. Windows, 9. Ultrasonic signal generator, 10. Frequency counter, 11. Oscilloscope, 12. Condencer microphone, 13. Microphone amplifier, 14. Measurement amplifier, 15. Lock-in amplifier, 16. Deviation amplifier, 17. Window comparator, 18. Gate circuit, 19. Absolute value circuit, 20. Voltage controlled oscillator, 21. Position indicator, 22. Overshoot alarm, 23. Stepping motor power supply and controller, 24. Stepping motor, 25. Drive mechanism, 26. CCD camera, 27. Videographic system board, 28. Color video monitor, 29. Microcomputer, 30. Data logger, 31. Thermocouple.

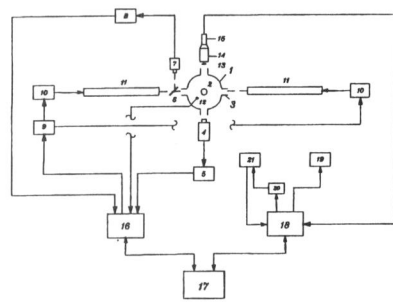

Figure 1-2. Heating and temperature control and image sampling subsystems

1. Resonance chamber, 2. Levitated particle, 3. Windows, 4. Infrared detector, 5. Indicator and transducer of infrared thermometer, 6. CO_2 laser beam splitter, 7. Detector of laser power meter, 8. Indicator and transducer of laser power meter, 9. Trigger-selector switch, 10. Neon bulb transformer, 11. CW CO_2 laser, 12. Thermocouple, 13. Fixed-focus lens, 14. Zoom lens, 15. CCD camera, 16. Data logger, 17. Microcomputer, 18. Videographic system board, 19. Color video monitor, 20. Color TV encoder, 21. Video recorder.

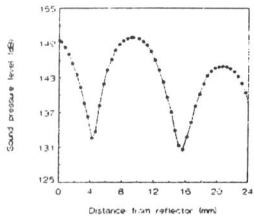

Figure 1-3. Axial pressure distribution above particle levitated at lowest sound pressure node

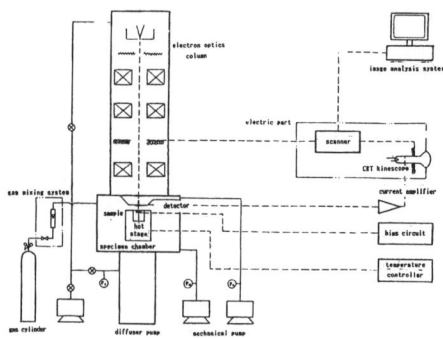

Figure 1-4. Diagrammatic sketch of the KYKY 1500 ESEM with hot stage

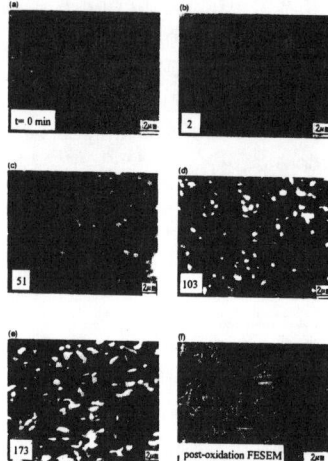

Evolution of iron oxidation at 500°C and P_{O2} = 3.2 Pa in the ESEM. (a) Unoxidized surface at 14°C; (b) t = 2 min; (c) t = 51 min; (d) t = 103 min; (e) t = 173 min (f) oxide whisker as shown by the AMRAY 1910 FESEM at 14°C, same conditions as (e).

Figure 1-5. Oxidation of Fe at 500° and P_O=3.2 Pa
to Whiskers

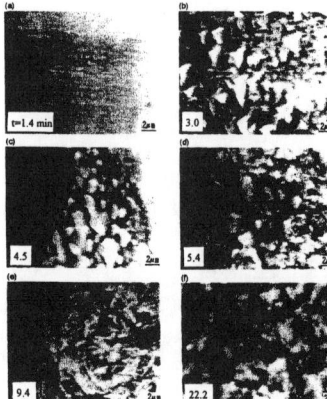

Evolution of iron oxidation at 700°C and P_{O2} = 44 Pa in the ESEM. (a) t = 1.4 min; (b) t = 3.0 min; (c) t = 4.5 min; (d) t = 5.4 min; (e) t = 9.4 min; (f) t = 22.2 min.

Figure 1-6. Oxidation of Fe at 700° and P_O=44 Pa
to Whiskers and Whisker Agglomerates

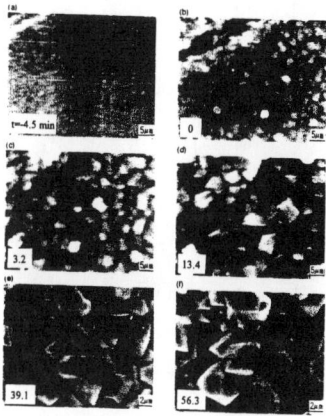

Evolution of iron oxidation at 600°C and P_{O2} from atmospheric air to $8.5 \cdot 10^{-4}$ Pa in ESEM. (a) $= -4.5$ min (547°C); (b) $t = 0$ min; (c) $t = 3.2$ min; (d) $t = 13.4$ min; (e) $t = 39.1$ min; (f) $t = 56.3$ min.

Figure 1-7. Oxidation of Fe at 600° and P_O=200/8.5x10^{-4} Pa to *Rectangular* and *Hexagonal Grains*.

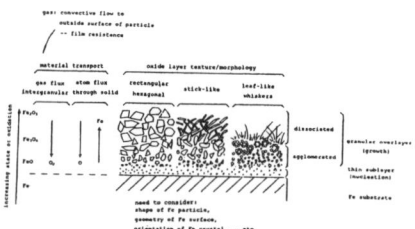

Figure 1-8. Plausible physical model for oxidation of Fe based on ESEM observations

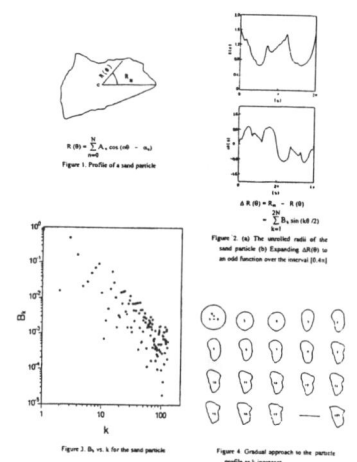

Figure 1-9. Fourier sine series for particle designation

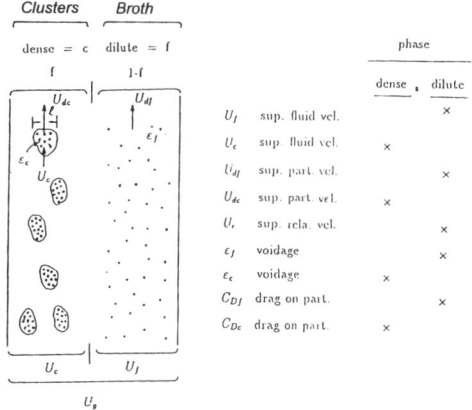

Figure 2-1. Physical Model for Multi-Scale Modeling of Particle-Fluid System

DISCRIMINATION

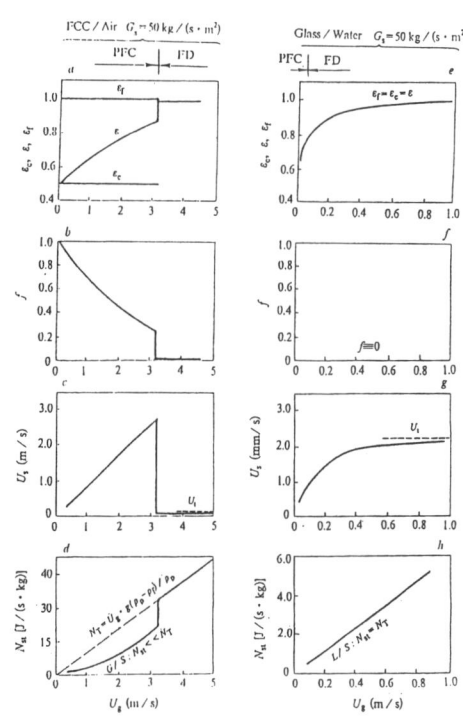

Figure 2-2. Differences between G/S and L/S Systems

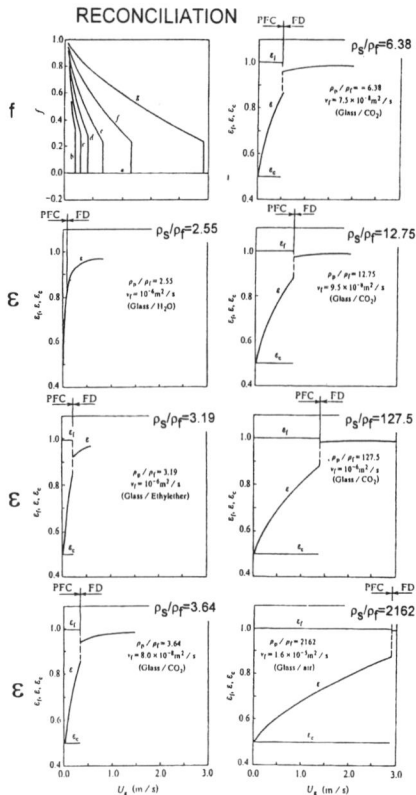

Figure 2-3. Gradual Transition from Particulate to Aggregative Fluidization

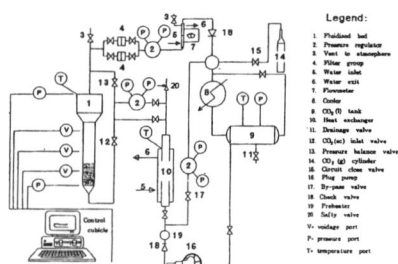

Figure 2-4. Apparatus for fluidization with pressurized CO_2

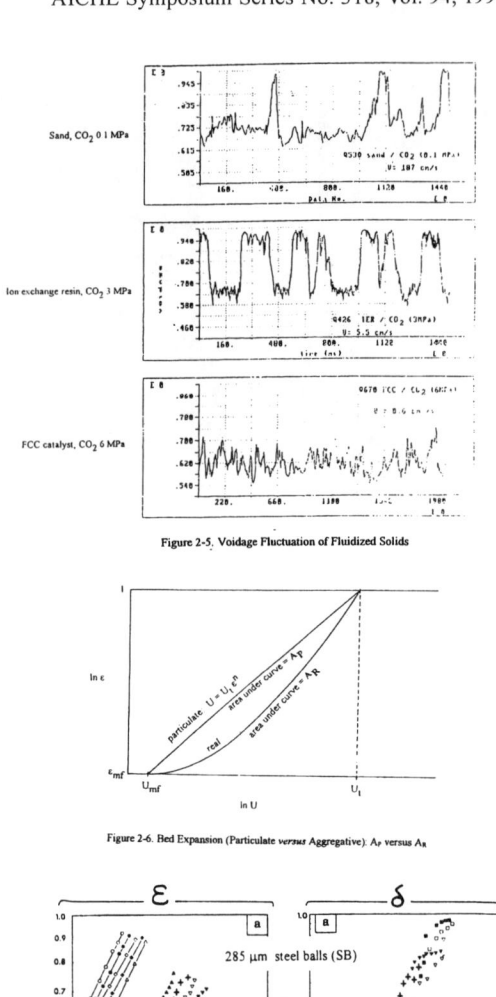

Figure 2-5. Voidage Fluctuation of Fluidized Solids

Figure 2-6. Bed Expansion (Particulate versus Aggregative): A_P versus A_R

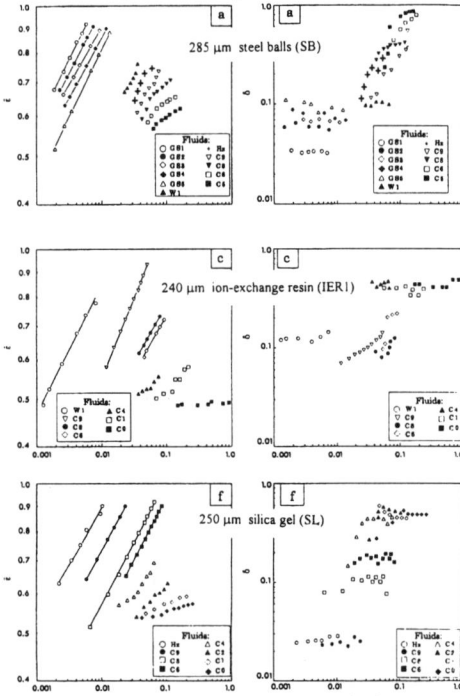

Figure 2-7. Bed Expansion and Local Heterogeneity

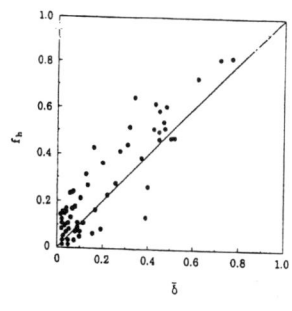

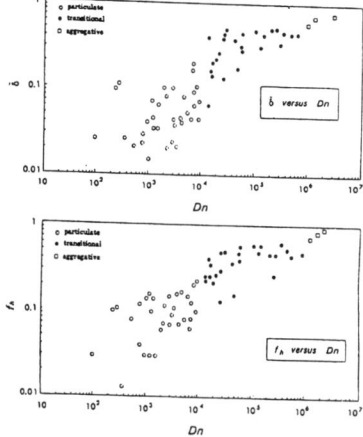

Figure 2-8. Interrelationship between $\bar{\delta}$, f_b and D_n

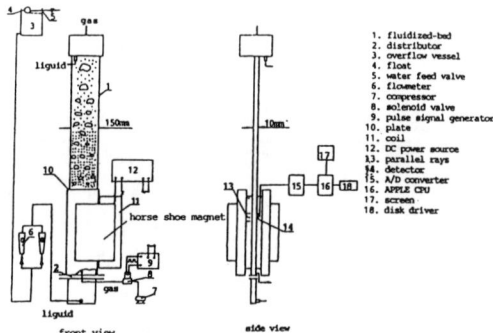

Figure 3-1. 2-D Magnetofluidization Apparatus

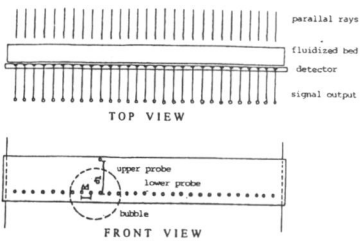

Figure 3-2. Photo-cell Array for Bubble Measurement

Figure 3-3. Computer Printout Showing Bubble Shape

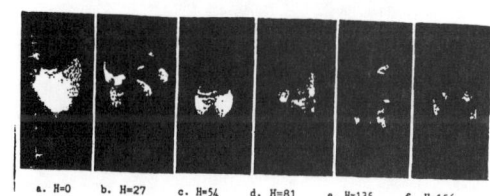

Figure 3-4. Bubble Behavior with Increasing Magnetic Field Strength

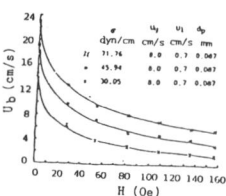

Figure 3-5. Effect of Magnetic Field Strength on Bubble Velocity

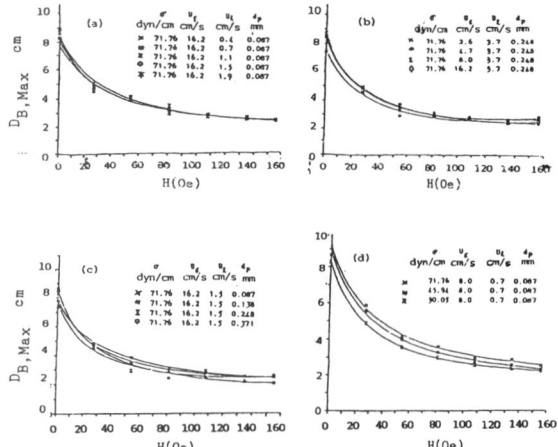

Figure 3-6. Effect of Magnetic Field Strength on Bubble Diameter

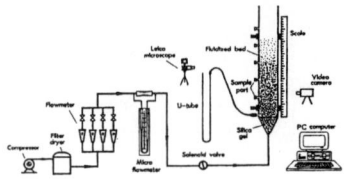

Figure 4-1. Experimental Fluidized Bed for Fine Powders

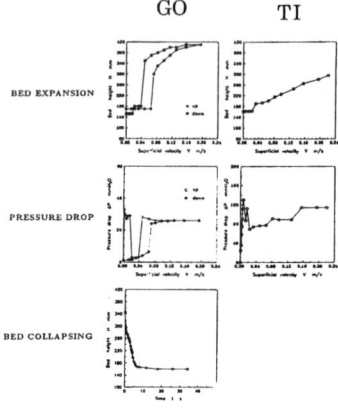

Figure 4-2. Bed Expansion, Pressure Drop and Bed Collapsing Tests for Fine Powders

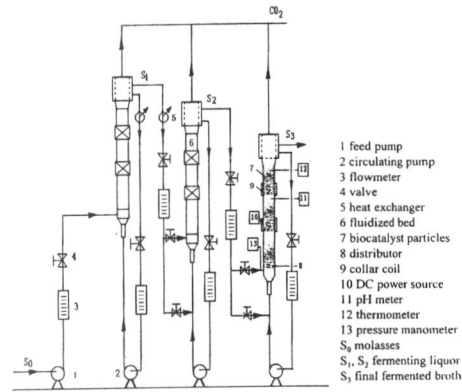

Figure 3-7. Continuous Magnetofluidized G/L/S Ethanol Bioreactor

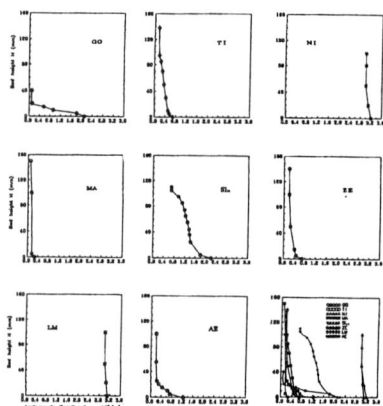

Figure 4-3. Variation of Agglomerate Size with Bed Height

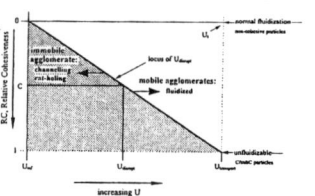

Figure 4-4. Comparison of fluidization behaviors for normal discrete particles and particle agglomerates

The Influence of a Particle Size Distribution on the Granular Dynamics of Dense Gas-Fluidized Beds: A Computer Simulation Study

B.P.B. Hoomans, J.A.M. Kuipers and W.P.M. van Swaaij
Department of Chemical Engineering, Twente University of Technology
PO Box 217, 7500 AE Enschede, The Netherlands

A hard-sphere discrete particle model of a gas-fluidized bed was extended in order to allow for a continuous particle size distribution to be taken into account. For each solid particle the Newtonian equations of motion are solved taking into account the inter-particle and particle-wall collisions. The gas-phase hydrodynamics is described by the spatially averaged Navier-Stokes equations for two-phase flow. Pressure peaks inside a slugging fluidized bed decreased with increasing geometric standard deviation of the log-normal size distribution. This could be observed during the initial stages of a simulation with 2400 particles starting from minimum fluidization conditions. In a bubble formation simulation using 40,000 particles of uniform size, small 'satellite' bubbles appeared above and alongside the main bubble. This could neither be observed in the simulation with polydispersed particles nor in the experiment duplicating the simulation which indicates the importance of taking a particle size distribution into account, especially when locally very close packing can prevail.

Due to increasing computer power granular dynamics simulations have become a very useful and versatile research tool to study the hydrodynamics of gas-fluidized beds. In these simulations the Newtonian equations of motion for each individual particle in the system are solved. Particle-particle and particle-wall interactions are taken into account directly which is a clear advantage over two-fluid models which require closure relations for the solids-phase stress tensor (Gidaspow (1), Sinclair and Jackson (2) and Kuipers et al. (3) among others). Tsuji et al. (4) developed a soft-sphere discrete particle model based on the work of Cundall and Strack (5). In their approach the particles are allowed to overlap slightly and from this overlap the contact forces are calculated subsequently. Hoomans et al. (6) used a hard-sphere approach in their discrete particle model which implies that the particles interact through binary, quasi-instantaneous, inelastic collisions with friction. An important advantage of a discrete particle model is given by the fact that a particle size distribution can easily be taken into account which is far more complex, if not impossible, in case a continuum modelling approach is adopted. Furthermore when simulating the particle dynamics directly with a uniform particle size the local particle configuration can approach a closest packing which leads to very low void fractions which in turn can influence the bed dynamics significantly. In this work the influence of a particle size distribution on the granular dynamics of dense gas-fluidized beds will be studied.

PARTICLE DYNAMICS

Since most details of the model are presented in a previous paper (Hoomans et al. (6)), the key features will be summarized briefly here. The collision model as originally developed by Wang and Mason (7) is used to describe a binary, instantaneous, inelastic collision with friction. The key parameters of the model are the coefficient of restitution ($0 \leq e \leq 1$) and the coefficient of friction ($\mu \geq 0$). In our hard-sphere approach a sequence of binary collisions is processed one collision at a time. This implies that a collision list is compiled in which for each particle a collision partner and a corresponding collision time is stored. A constant time step is used to take the external forces into account and *within* this time step the prevailing collisions are processed sequentially. In order to reduce the required CPU time neighborlists are used in order to decrease the number of particles to be checked for possible collisions. Efficient algorithms obtained from the field of Molecular Dynamics (MD) are employed to achieve very efficient computational procedures.

External Forces

The incorporation of external forces differs somewhat from the approached followed in our previous paper (Hoomans et al. (6)). In this work we use the external forces analogous to those implemented in the two-fluid model described by Kuipers et al. (3) where, of course, the forces now act on a single particle:

$$m_p \frac{d\mathbf{v}_p}{dt} = m_p \mathbf{g} + \frac{V_p \beta}{(1-\varepsilon)}(\mathbf{u} - \mathbf{v}_p) - V_p \nabla p \quad (1)$$

where m_p represents the mass of a particle, $\mathbf{v}_p$ its velocity, $\mathbf{u}$ the local gas velocity and V_p the volume of a particle. In Equation (1) the first term is due to gravity and the third term is the force due to the pressure gradient. The second term is due to the drag force where β represents an interphase momentum exchange coefficient as it usually appears in two-fluid models. For low void fractions ($\varepsilon < 0.80$) β is obtained from the well-known Ergun equation:

$$\beta = 150 \frac{(1-\varepsilon)^2}{\varepsilon} \frac{\mu_g}{d_p^2} + 1.75(1-\varepsilon)\frac{\rho_g}{d_p}|\mathbf{u} - \mathbf{v}_p| \quad (2)$$

where d_p represents the particle diameter, μ_g the viscosity of the gas and ρ_g the density of the gas. For high void fractions ($\varepsilon \geq 0.80$) the following expression for the interphase momentum transfer coefficient has been used:

$$\beta = \frac{3}{4} C_d \frac{\varepsilon(1-\varepsilon)}{d_p} \rho_g |\mathbf{u} - \mathbf{v}_p| \varepsilon^{-2.65} \quad (3)$$

The drag coefficient C_d is a function of the particle Reynolds number:

$$C_d = \begin{cases} \frac{24}{\text{Re}_p}(1 + 0.15\,\text{Re}_p^{0.687}) & \text{Re}_p < 1000 \\ 0.44 & \text{Re}_p \geq 1000 \end{cases} \quad (4)$$

where the particle Reynolds number in this case is defined as follows:

$$\text{Re}_p = \frac{\varepsilon \rho_g |\mathbf{u} - \mathbf{v}_p| d_p}{\mu_g} \quad (5)$$

For the integration of Equation (1) a simple explicit first order scheme was used to update the velocities and positions of the particles.

PARTICLE SIZE DISTRIBUTION

Hardly any distribution of particle size encountered in fluidization studies is symmetrical, most of them are skewed to larger diameters (Seville et al. (8)). A symmetrical size distribution like a normal or a Gaussian distribution is therefore not representative for particles used in laboratory or industrial practice. In this work the particle diameters are obtained from a log-normal distribution which is asymmetrical and can be represented in mathematical terms as follows:

$$df = \frac{1}{\sqrt{2\pi} d_p \ln \sigma_g} \exp\left[-\frac{(\ln d_p - \ln d_{p,CMD})^2}{2(\ln \sigma_g)^2}\right] dd_p, \quad (6)$$

where df is the fraction of particles having diameters whose logarithms lie between $\ln d_p$ and $\ln d_p + d(\ln d_p)$. In Equation (6) $d_{p,CMD}$ is the count median diameter and σ_g is the geometric standard deviation. When creating the particle size distribution all the diameters which are smaller than $d_{p,CMD} - \sigma_g$ or larger than $d_{p,CMD} - \sigma_g$ are rejected which mimics the effects of sieving. An example of a particle size distribution generated with this method is represented in a (discrete) frequency histogram in Figure 1.

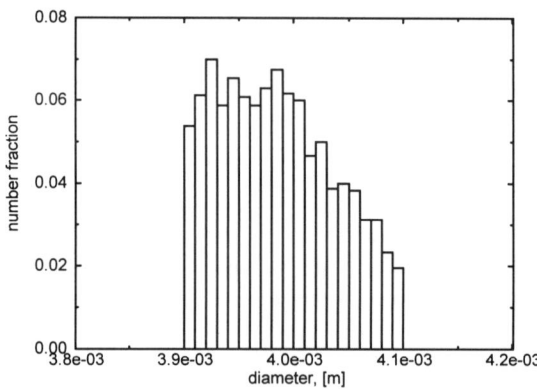

Figure 1, Frequency histogram of a log-normal particle size distribution with a count median diameter of 4.0 [mm] and a σ_g of 0.1 [mm].

GAS PHASE HYDRODYNAMICS

The motion of the gas-phase is calculated from the following set of equations which can be seen as a generalised form of the spatially averaged Navier-Stokes equations for a two-phase gas-solid mixture (Kuipers et al. (3)).

Continuity equation gas phase:

$$\frac{\partial(\varepsilon \rho_g)}{\partial t} + (\nabla \cdot \varepsilon \rho_g \mathbf{u}) = 0 \ . \quad (7)$$

Momentum equation gas phase:

$$\frac{\partial(\varepsilon\rho_g \mathbf{u})}{\partial t} + (\nabla \cdot \varepsilon\rho_g \mathbf{u}\mathbf{u}) = -\varepsilon\nabla p - \mathbf{S}_p - (\nabla \cdot \varepsilon\tau_g) + \varepsilon\rho_g \mathbf{g} \quad (8)$$

In this work transient, two-dimensional, isothermal (T = 293 K) flow of air at atmospheric conditions is considered. The constitutive equations and the boundary conditions used can be found in Hoomans et al. (6). The void fraction (ε) is calculated from the particle positions in the bed. There is one important modification with respect to our previous model and that deals with the way in which the two-way coupling between the gas-phase and the particle motion is established. In the present model the reaction force to the drag force exerted on a particle per unit of volume is fed back to the gas-phase through the source term $\mathbf{S}_p$ [Nm^{-3}]. A more detailed discussion on this approach can be found in Delnoij et al. (9).

RESULTS

Influence of Distribution Width

At first the influence of the distribution width was studied. Simulations were performed using 2400 particles with a count median average diameter of 4.0 [mm] and a density of 2700 kg/m^3 (u_{mf} = 1.78 [m/s]) contained in a system of 0.15 [m] width and 0.5 [m] hight. A discretization of 15 cells horizontally and 25 cells vertically was applied. A time step of 10^{-4} [s] was used and all simulations were run for 10 [s] real time. The coefficient of restitution (e) was set equal to 0.9 and the coefficient of friction (μ) was set equal to 0.3 for both particle-particle and particle-wall collisions. The initial configurations were obtained by placing the particles in the system and allowing them to fall under the influence of gravity while the gas inflow was set equal to u_{mf}. Simulations were performed for a system consisting of particles of uniform size and systems consisting of particles with a log-normal size distribution with a geometric standard deviation σ_g = 0.1, 0.5 and 1.0 [mm] respectively. A homogeneous gas inflow at 1.5 u_{mf} was specified at the bottom of the system.

In Figure 2 the pressure fluctuations inside the bed at 0.2 [m] above the centre of the bottom plate are presented for all the four cases. It can be observed that the pressure peaks decrease with increasing geometric standard deviation of the size distribution. This can be explained by the lower void fraction in the uniform case due to the closer packing which results in a higher force acting on the particles which in turn causes the higher pressure peaks. After the first 1.5 [s] the differences become far less pronounced which indicates that especially in situations where close packing can occur such as at minimum fluidization conditions it is important to take polydispersity into account.

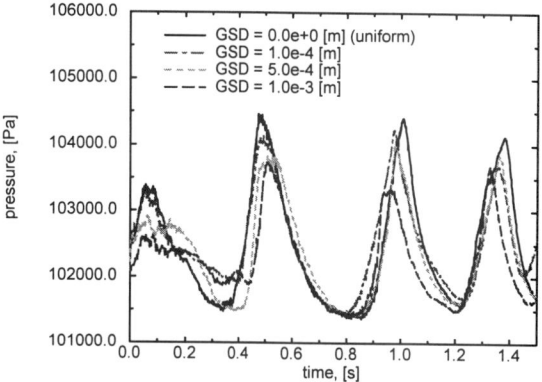

Figure 2, Pressure fluctuations 0.2 [m] above the centre of the bottom plate as a function of time for the four cases, GSD = geometric standard deviation (σ_g).

Experimental Validation

In order to test whether the simulation results were improved by taking the particle size distribution into account a comparison with experiment was performed. The main features of the experimental set-up were reported earlier (Hoomans et al. (6)). A sieve fraction of glass ballotini particles between 800 and 900 [µm] with a density of 2930 [kg/m^3] (u_{mf} = 0.5 [m/s]) was used. The bed (width 0.2 [m] and hight 0.3 [m]) was equipped with a porous bottom plate which featured a central nozzle (15 [mm]) through which excess gas could be injected into the bed. The background fluidization velocity was kept equal to u_{mf} whereas during the first 0.2 [s] excess gas was injected through the nozzle at 5 u_{mf}. The simulations were performed using 40,000 particles. The coefficient of restitution was 0.96 for particle-particle collisions and 0.86 for particle-wall collisions and the coefficients of friction were 0.15 in both cases. A discretization of 39 by 60 cells was used together with a time step of 10^{-4} [s]. A simulation was performed where the particle diameters were obtained from a log-normal distribution with a count median average of 850 [μm] and a geometric standard deviation of 50 [μm] as well as a reference simulation with particles of uniform diameter (850 [μm]). Snapshots at t = 0.2 [s] of both simulations and the experiment are presented in Figure 3.

It can be observed that in the case of the uniform particle assembly small 'satellite' bubbles appear above an alongside the main bubble. This is neither observed in the experiment nor in the simulation with the polydisperse particle assembly. These 'satellite' bubbles are probably due to low local void fraction due to close packing. The size of the main bubble agrees rather well with the experiment for both simulations which is rather encouraging especially since all model parameters were obtained on beforehand and independently. Further improvement can be achieved by extending the model to three dimensions because although the experimental bed was quasi two-dimensional it was still 16 [mm] deep.

CONCLUSIONS

Granular dynamics simulations have been performed with particles which diameters have been obtained from a log-normal distribution. The main influence of the polydispersity could be observed when the particles were in a rather close packing. Pressure peaks inside a slugging fluidized bed decreased with increasing geometric standard deviation of the log-normal size distribution during the initial stages of a simulation starting from minimum fluidization conditions. In a bubble formation simulation with particles of uniform size small 'satellite' bubbles appeared above and alongside the main bubble. This could neither be observed in the experiment nor in the simulation with the polydisperse particles. This indicates that it is especially important to take a particle size distribution into account in systems where locally very dense particle configurations can occur.

Due to the fact that a particle size distribution can be dealt with in a natural and fundamental manner (i.e. without invoking difficulties in formulating closure laws for particle stress) it is anticipated that the effect of fines on fluidization behaviour, which is well known to the experimentalist but which is unfortunately poorly understood, can now be studied in detail.

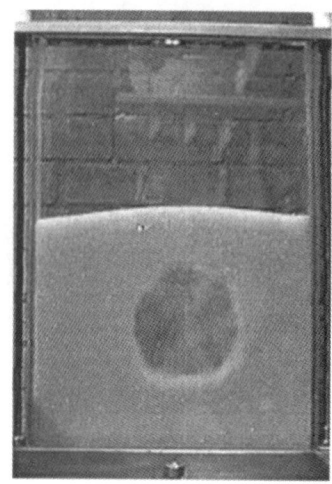

a)

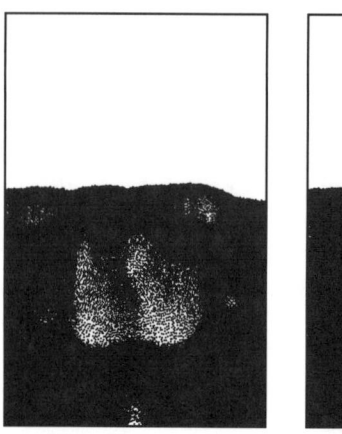 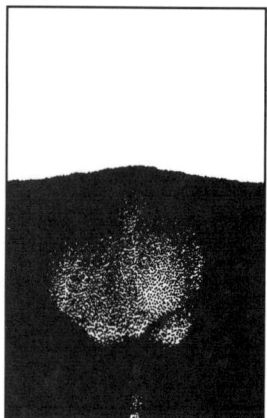

b)   c)

Figure 3, Snapshots at t = 0.2 [s] of: a) experiment, b) simulation with uniform particles, c) simulation with log-normal particle size distribution.

NOTATION

- C_d drag coefficient, [-]
- e coefficient of restitution, [-]
- d_p particle diameter, [m]
- g gravitational acceleration, [m/s^2]
- m_p particle mass, [kg]
- p pressure, [Pa]
- $\mathbf{r}$ position vector, [m]
- $\mathbf{S}_p$ momentum source term Eq. (8), [N/m^3]
- t time, [s]
- $\mathbf{u}$ gas velocity vector, [m/s]
- $\mathbf{v}_p$ particle velocity vector, [m/s]
- V_p particle volume, [m^3]

Greek symbols

- β defined in Eqs (2) and (3), [kg/m^{3}s]
- ε void fraction, [-]
- μ coefficient of friction, [-]
- μ_g gas viscosity, [kg/ms]
- τ gas-phase stress tensor, [kg/ms^2]
- ρ_g gas density, [kg/m^3]
- σ_g geometric standard deviation (GSD), [m]

LITERATURE CITED

1. Gidaspow, D., *Multiphase Flow and Fluidization*, Academic Press, Boston (1994).

2. Sinclair, J. L. and R. Jackson "Gas-particle flow in a vertical pipe with particle-particle interactions," *AIChE J.*, **35**, 1473 (1989).

3. Kuipers, J. A. M., K. J. van Duin, F. P. H. van Beckum and W. P. M. van Swaaij, "A numerical model of gas-fluidized beds," *Chem. Engng Sci.*, **47**, 1913 (1992).

4. Tsuji, Y., T. Kawaguchi, and T. Tanaka, "Discrete particle simulation of two dimensional fluidized bed," *Powder Technol.* 77, 79 (1993).

5. Cundall P. A. and O. D. L. Strack, "A discrete numerical model for granular assemblies," *Géotechnique*, **29**, 47 (1979).

6. Hoomans, B. P. B., J. A. M. Kuipers, W. J. Briels and W. P. M. van Swaaij, "Discrete particle simulation of bubble and slug formation in a two-dimensional gas-fluidised bed: a hard-sphere approach," *Chem. Engng Sci.*, **51**, 99 (1996).

7. Wang Y. and M. T. Mason, "Two-dimensional rigid-body collisions with friction," *J. Appl. Mech.*, **59**, 635 (1992).

8. Seville, J. P. K., U. Tüzün, and R. Clift, *Processing of Particulate Solids*, Blackie Academic & Professional (1997).

9. Delnoij, E., F. A. Lammers, J. A. M. Kuipers and W. P. M. van Swaaij, "Dynamic simulation of dispersed gas-liquid two-phase flow using a discrete bubble model," *Chem. Engng Sci.*, **52**, 1429 (1997).

Determination of Lateral Dispersion Coefficients in the Dilute Region of Fluidized Beds

Matthew R. Hyre
LeTourneau University, Longview, TX

Leon R. Glicksman
Massachusetts Institute of Technology, Cambridge, MA

The dynamics of ciculating fluidized bed solids dispersion has been a subject of intensive study over the past decade as the importance and use of fluidiation techniques have increased. An essential element in that process is the way in which particles are dispersed throughout the bulk of the fluid under the action of turbulent and interparticle collisional forces which are random in both space and time. A Lagrangian model has developed to study radial particle dispersion and deposition in the dilute core of a CFB. The model includes the effects of particle inertia, crossing-trajectories, bed voidage, and particle-particle interactions. Particles are tracked through the turbulent flow field using a particle/eddy interaction time to describe the effect of the interaction between the particles and fluid, and using a collisional timescale to describe the effect of particle collisions. Calculated dispersion coefficients are compared to experimentally measured coefficients in CFB's, FCC reactors, and dilute pneumatic transport multiphase flows. Results show good agreement between predicted and measured dispersion coefficients.

INTRODUCTION

One of the cited advantages of CFB's is their inherently high degree of mixing. Effectiveness of chemical processes, such as combustion of low heating value waste fuels for which CFB's are used, depends on the ability to effectively mix separate streams of reactants. However, focus on local mixing details has been bypassed in most CFB mixing studies. Instead, most of the models for particle mixing look at fairly large time and length scales, for instance, the mean mixing time for a particle to traverse the entire reactor. In addition, homogeneity of the flow in axial and/or radial directions is usually assumed. This approach allows only an overall average view of the mixing process. Study of the effects of mixing on fuel devolatilization and volatile combustion, a process with a timescale of roughly a second, must take local mixing into account. Furthermore, it is the combination of local mixing phenomena which result in overall mixing. An understanding of the mechanisms contributing to mixing, and prediction of mixing characteristics of shorter time and length scales are essential to a complete understanding of CFB mixing. Therefore, this studied focused on the development of a particle dispersion model capable of predicting the solids dispersion rate in the core of a CFB for timescales on the on the order of time required for fuel devolatilization.

MODEL DESCRIPTION

A Lagrangian technique was used to predict the lateral solids dispersion in the core of a CFB. The simulation included the effects of particle inertia, 'crossing-trajectories', particle sphericity, bed voidage, and particle-particle interactions. Ideally, this approach requires knowledge of the full time history of the turbulent flow, obtained by solving the instantaneous (unaveraged) Navier-Stokes equations. Since this is not realistic, the turbulence was simulated as a stochastic process, where mean values are determined from the time-averaged Navier-Stokes solutions. Each eddy is then characterized by a mean time and length scale, and the velocity fluctuations are randomly-generated in a Gaussian manner as a particle enters an eddy. The result is a Monte-Carlo procedure where many particle trajectories (realizations) must be computed to obtain averaged properties.

The difficulty in the Lagrangian approach is choosing appropriate time and length scales of the fluid turbulence. Approximations for the time and length scales for the fluid turbulence were determined from data provided by Laufer [1] for flow in pipes. This data

---

M. R. Hyre is now with Black and Decker, Windsor, Connecticut.

was used to correlate the ratio of eddy length to pipe diameter versus the pipe Reynolds number. A modified form of the turbulence modulation model of Yuan and Michaelides [2] was used to correct the gas turbulent energy for the presence of a relatively dilute particle loading [3].

For large or heavy particles, the interaction time between a particle and an eddy is controlled by the time it takes a particle to fall through the eddy. This is the well known "crossing trajectories" effect. For small particles, the interaction time is controlled by the horizontal crossing time of the particles or the mean decay time of the eddy. For particles typical of CFB's, the crossing trajectories timescale is almost always smaller than the eddy decay time. Therefore, for very dilute regions in the core, the crossing trajectories timescale limits the eddy/particle interaction time. The drag force expression of Turton and Levenspiel [4], modified for the effects of particle sphericity and bed core voidage as suggested by Ganser [5] and Barnea and Mizrahi [6], was used to describe the eddy/particle drag interaction.

Including collisions in the analysis introduces the possibility that the particle-eddy interaction time is no longer controlled by either the horizontal or vertical particle-eddy crossing trajectories, or by the eddy decay time. This is especially true in regions of higher densities. Rather, particles within an eddy may collide with one another due to both the distribution in radial fluctuating velocities and the distribution in axial slip velocities of different particle sizes. During the time the particles are in the eddy, they gain a component of velocity due to the drag force in the direction of the eddy gas velocity, and also due to collisions with other particles. Expressions for the various timescales which control particle-eddy and particle-particle interactions are given in Table 1.

REQUIRED MODEL INPUT

The following variables are required input to the model.

1. Bed Diameter
2. Bed Height
3. Mean Particle Diameter
4. Standard Deviation of Particle Size Distribution (default equal to the mean particle diameter)
5. Particle Sphericity (default equal to 0.8)
6. Particle Density
7. Bed Temperature
8. Bed Pressure
9. Gas Superficial Velocity
10. Solids Recycle Rate
11. Particle/Particle Coefficient of Restitution (default equals 1)
12. Exit Collection Efficiency (default equals 0.8)
13. Number of Particles to Simulate
14. Number of Interactions per Particle to Simulate

Table 1: Expressions for Various Time Scales

Time Scale	Equation	Region where Time Scale Will be Controlling
Particle-Eddy Vertical Crossing Time	$\dfrac{l_e}{u_t}$	Large particles, relatively dilute suspensions
Particle-Eddy Horizontal Crossing Time	$\dfrac{l_e}{u_t}\left[\dfrac{\exp\left(\frac{3}{8}C_D\frac{l_e}{R}\frac{D}{d_p}\frac{\rho_f}{\rho_s}\right)-1}{\frac{3}{8}C_D\frac{l_e}{R}\frac{D}{d_p}\frac{\rho_f}{\rho_s}\left(1\pm\frac{v_{p_o}}{u_e}\right)}\right]$	Small particles, relatively dilute suspensions
Eddy Decay Time	$1.6\left(\dfrac{l_e}{u_e}\right)$	Very large diameter beds, or very small particles, dilute suspensions
Time Between Collisions - Relatively Dilute Suspensions	$\tau_{c_{ij}} = \dfrac{1}{n_j \pi d_{ij}^2 (u_{t_1} - u_{t_2})}$	Relatively dilute suspensions, large particle size distributions
Time Between Collisions - Dense Suspensions	$\tau_{c_{ij}} = \dfrac{1}{n_j \pi d_{ij}^2 v_{p_r}'}$	Very dense suspensions

PROCEDURE FOR THE DETERMINATION OF THE LATERAL SOLIDS DISPERSION COEFFICIENT

The simulation to determine the lateral solids dispersion coefficient is initialized by selecting a particle diameter. The particle diameter is taken from a Gaussian distribution with a mean equal to the input mean particle diameter and a standard deviation based on the input particle size distribution standard deviation. The mean eddy length and eddy speed are then calculated from the model based on the data of Hutchinson *et al.* [7].

At this point, the particle-eddy interactions begin for the chosen particle. For each interaction, an eddy length is selected from a Gaussian distribution with a mean equal to the length determined from the data of Laufer and a standard deviation equal to one half the mean. An eddy speed is also determined from a Gaussian distribution with a mean equal to zero and a r.m.s. value equal to the eddy speed modified for turbulence damping due to the presence of particles. The eddy velocity (direction of speed) is determined by randomly multiplying the speed by either 1 or -1. Finally, the component of particle velocity due to collisions is determined. The mean collisional component of the radial particle fluctuating velocity is determined using the particle collision model of Kumaran and Koch [8,9], and a particle diameter ratio equal to the ratio of the current particle diameter to the previous particle diameter. The direction of the collisional component of the particle fluctuating velocity is determined by randomly multiplying the r.m.s. value by either 1 or -1. For the initial dispersion coefficient calculation, the collisional component is obviously zero.

For each interaction, the interaction time is determined from the minimum of the vertical particle-eddy crossing time, horizontal particle-eddy crossing time, eddy decay time, and the time between successive collisions. The time between successive collisions is determined from an input solid fraction or an estimated solid fraction profile developed from the previous value of lateral dispersion. For the initial calculation, the time between successive collisions is assumed to be infinite. After the interaction time is determined, the particle equation of motion is solved using the final velocity of the previous interaction as the starting velocity. For the initial interaction, the particle velocity at the beginning of the interaction is assumed to be zero (the particle velocity at the beginning of the initial interaction has no influence on the final results if more than 100 interactions are used).

After the particle has proceeded through the user specified number of interactions, another particle is selected based on the Gaussian distribution; the previous steps are repeated. This process continues until the total number of particles specified by the user has been tracked through the interaction process.

After all the particles have been tracked through the particle-eddy interaction process, the average mean square displacements, velocities, interaction times, and dispersion coefficients, along with all other pertinent data is evaluated.

COMPARISON OF MODEL RESULTS TO EXPERIMENTAL DATA

Lateral Dispersion Coefficients in Circulating Fluidized Beds

Table 2 presents sources of data currently available on lateral dispersion coefficients in the dilute region of CFB's or FCC reactors.

Table 2: Lateral Solid Dispersion Coefficient Measurements in CFB's

Researcher	D (cm)	d_p (μm)	ρ_s (kg/m^3)	u_o (m/s)	ε_{p_r} (cm^2/s)	Pe_r
Westphalen and Glicksman [10]	20.5	180	2350	3 - 5.5	0.4 - 8.6	412 - 10000
van Zoonen [11]	5	65	1600	2.5 - 12	0.94 - 12.0	250 - 833
Wei et al. [12]	14	54	1710	2.3 - 9	12 - 50	70 - 300
Koenigsdorff and Werther [13]	20	60	3217	3 - 4	18 - 47	150 - 400

Figures 1 through 4 present comparisons between the predicted and experimentally determined lateral solids dispersion coefficients. In all cases, the present model is in good agreement with the experimental data. Figures 1 through 4 indicate that the dispersion model is able to predict effective lateral solids dispersion coefficients over a wide range of conditions: from the standard cold CFB models (Westphalen and Glicksman, [10]; Koenigsdorff and Werther [13]), to large scale FCC reactors (van Zoonen [11]).

Figure 1
Comparison of Model with Dispersion Data of Westphalen and Glicksman (1995)

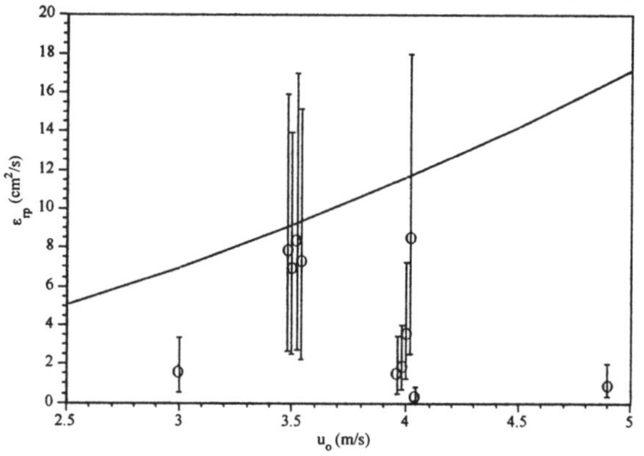

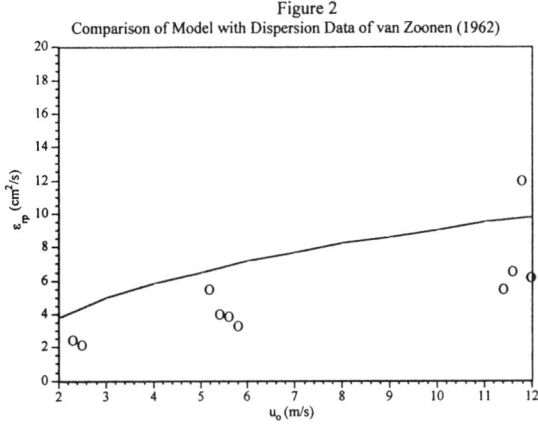

Figure 2
Comparison of Model with Dispersion Data of van Zoonen (1962)

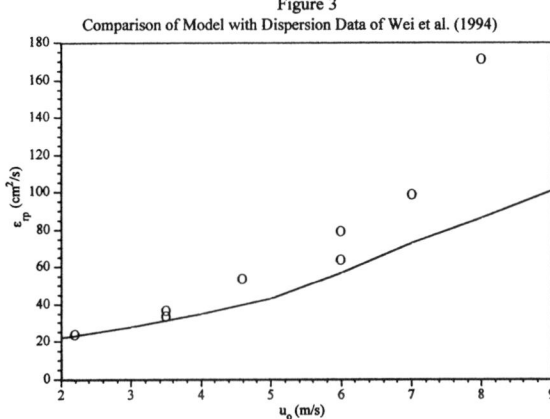

Figure 3
Comparison of Model with Dispersion Data of Wei et al. (1994)

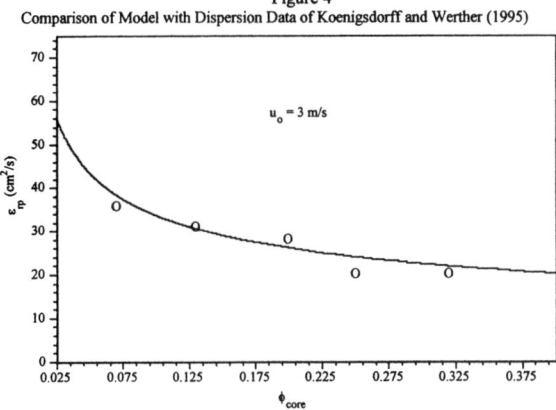

Figure 4
Comparison of Model with Dispersion Data of Koenigsdorff and Werther (1995)

It should be noted that for the van Zoonen [11] flow configuration, the "crossing-trajectories" effect is not significant since the average time it takes for an eddy to decay is less than the time it takes for an average particle to fall through the eddy. The good agreement between the model and the experimental data suggest that the model accurately distinguishes between the crossing-trajectories and eddy decay effects.

Lateral Dispersion Coefficients in Dilute Pneumatic Transport

Table 3 shows a comparison of the model against the pneumatic transport data of Lee et al. [14]. The model should be especially good at predicting this data since the flow configuration was for very dilute pneumatic transport of droplets. The most inaccurate aspect of the proposed model is the first order correction for the inter-particle collisions. Since the time between collisions was very long compared to the droplet relaxation time in the flow configuration of Lee et al., this inaccuracy is eliminated. Agreement between the experimental data and the numeric model is very good, especially for the 50 and 150 micron droplets.

Table 3: Comparison of Model with Data of Lee et al. [14]

d_p (microns)	Re_D	Measured v'_{p_r} (cm/s)	Model v'_{p_r} (cm/s)	Measured ε_{p_r} (cm^2/s)	Model ε_{p_r} (cm^2/s)
50	36000	32.6	34.2	9.2	10.9
50	52000	46.6	40.2	13.5	12.5
90	36000	23.9	23.6	8.2	6.3
90	52000	27.7	28.3	10.5	7.8
150	36000	18.4	18.0	4.1	4.3
150	52000	21.2	22.0	6.5	5.9

At a droplet diameter of 50 microns, the eddy decay time is the controlling timescale for particle-eddy interaction. At 150 microns, the particle-eddy interaction time is controlled by the time it takes a droplet to cross the eddy. At a droplet diameter of 90 microns, the eddy decay time is nearly equal to the particle-eddy vertical crossing time. This is the most likely source for the larger error at this diameter. Since the simulation uses different models depending on the smallest timescale, the inaccuracies in fluid turbulence timescales and the particle properties are magnified if the selected model is not the most appropriate.

NOTATION

C_D drag coefficient
D   bed diameter, m
d_{ij} mean diameter of colliding particles, m
d_p particle diameter, m
L   bed height, m
l_e eddy length, m
n_j number density of j particles, 1/m^3
Pe_r Peclet number based on lateral dispersion
R   bed radius, m

u_e eddy speed, m/s
u_o superficial velocity, m/s
$u_{p,r}$ radial component of fluctuating velocity, m/s
u_t terminal velocity

Greek Symbols

ε_{pr} lateral dispersion coefficient, m²/s
ρ_f gas density, kg/m³
ρ_s particle density, kg/m³
τ_c time between particle collisions, s

LITERATURE CITED

1. Laufer, J., *The Structure of Turbulence in Fully Developed Pipe Flow*, NACA Rep., Report 1174, (1954).

2. Yuan, Z., and E.E. Michaelides, *Int. J. Multiphase Flow*, **18**, p. 779, (1992).

3. Hyre, M., *Aspects of Hydrodynamics and Heat Transfer in Circulating Fluidized Beds*, Ph.D. Thesis, MIT, (1995).

4. Turton, R., and O. Levenspiel, *Powder Technol*, **47**, p. 83, (1986).

5. Ganser, G.H., *Powder Technol.*, **77**, p. 143, (1993).

6. Barnea, E. and J. Mizrahi, *Chem Eng. Jl.*, **5**, p. 171, (1973)

7. Hutchinson, P., F.G. Hewitt, and A.E. Dukler, *Chem. Engng. Sci.*, **26**, p. 419, (1971).

8. Kumaran, V., and D.L. Koch, *J. Fluid Mech.*, **247**, p. 643, (1993).

9. Kumaran, V., and D.L. Koch, *J. Fluid Mech.*, **247**, p. 623, (1993).

10. Westphalen, D., and L.R. Glicksman, *Powder Technol.*, **82**, p. 153, (1995).

11. Van Zoonen, D., *Proceedings of Symposium on Interaction between Fluids and Particles, London, Inst. Chemical Engineers*, London, p. 64, (1962).

12. Wei, F., Z. Wang, Y. Jin, Z. Yu, and W. Chen, *Powder Technol.*, **81**, p. 25, (1994).

13. Koenigsdorff, R., and J. Werther, *Powder Technol.*, **82**, p. 317, (1995).

14. Lee, M.M, T.J. Hanratty, and R.J. Adrian, *Int. J. Multiphase Flow*, **15**, p. 787, (1989).

Long-Range Connectivity in Slow-Shearing Granular Flows

J. Baxter and U. Tüzün
Department of Chemical and Process Engineering, University of Surrey,
Guildford, GU2 5XH, UK.

D.M. Heyes
Department of Chemistry, University of Surrey,
Guildford, GU2 5XH, UK.

Long-range connectivity, the co-operative behavior of granules at and beyond the lengthscale of the granule size, is a commonly-observed phenomenon of slow-shearing flows. We investigate long-range connectivity in the pouring of heaps, using discrete particle computer simulations and focusing our attention on the characteristics of various granule-granule interaction models which are the basis of simulation strategies. We find that explicit provision for long-range connectivity, in the form of a molecular-dynamics style soft-sphere potential model, is most appropriate in reproducing a range of experimental and theoretical observations on slow-shearing granular flows. The behavior of such a model is compared against a simple linear spring model, and found to reproduce the heterogeneous nature of stress and voidage distributions within the heap much more closely. The characteristic feature of the soft-sphere model is that it gives an extended range of interaction between granules; the range of influence of granules is deemed to extend some distance beyond their nominal boundaries. This is found to be essential for realistic simulation of slow-shearing granular flows.

The behaviour of granular media in many situations suggests a high degree of co-operation between the grains. For example, the failure of materials under bulk strain (for example, in a shear cell test) is concentrated into narrow failure zones that are only a few particle diameters across [1]. In quasi-static assemblies, such as heaps, formed by slow shearing flows, such as by pouring, there is considerable inhomogeneity in the stress distribution within the assembly. This is indicative of co-operative behaviour via the spontaneous organization of particles.

A number of experimental observations on granular heaps point to the importance of what we term long-range connectivity, the co-operative behaviour of grains over lengthscales much greater than a single particle diameter. In many cases, theoretical investigations have attempted to build upon the experimental observations. The present numerical study investigates certain aspects of poured granular heaps for which there is experimental and theoretical evidence in the literature; the profiles of granule-granule contact forces, and of local voidage, within the heap and the profile of contact forces on the supporting base.

The study involves discrete particle, Newtonian dynamics computer simulations. Our granular dynamics (GD) approach relies on models for the computation of particle-particle interaction forces, and the numerical integration of Newtonian laws of motion over small finite time-steps, to yield particle velocities and trajectories. In particular, we examine the effects of the particle-particle interaction model on the behaviour of poured heaps, in light of experiments and theory. For the purposes of this paper, we restrict our attention to assemblies of monosized two-dimensional disks, although we also consider heaps constructed from polydisperse particles, and study phenomena such as particle size segregation.

MODEL

The mathematical model we use here for GD simulations is described elsewhere [2] thus, only a brief account of the essential points is given here. The particle assembly evolves as a consequence of interactions between grains, which are assumed pairwise additive (each grain-grain contact evolves independently of the others). The forces and moments acting on granules at any point in time arise from elastic normal interactions, friction and gravity. Furthermore, energy dissipation (in the normal and tangential directions) is modeled in the form of a damping force. Of these five sources, in modeling terms the first two are the most significant. The work of Cundall and Strack [3] used a linear spring model for the elastic normal interaction. Later implementations of the GD approach, such as that used by Thornton and Yin [4] modified the approach to include classical (Hertzian) contact mechanics.

Alternatively, among others Ristow [5], Langston *et.al* [6] and ourselves [2][7] have used interaction potentials originating from the molecular dynamics (MD) approach. The linear spring model takes the form: $F(r)=k(\sigma-r)$ where k is the spring stiffness, r the centre-to-centre separation of the two particles, and σ the particle diameter. The term $(\sigma-r)$ represents the 'overlap' of the two particles. The Hertzian model is also dependent on the overlap; for this model, $F(r)=K(\sigma-r)^{3/2}$, where K is a group of constants including the elastic modulus of the particles. In the MD-type models, the force is not an explicit function of the overlap, but is nevertheless dependent on separation; in general, $F(r)=\phi(\sigma/r)^n$ where ϕ is a constant and n an index describing the 'hardness' of the interaction. There are some important differences between these functional forms.

The experience from previous studies is that the most significant characteristic of an interaction model is the separation at which interaction is deemed to cease. For the spring and Hertz models this is fairly obviously at a separation of $r=\sigma$, where the interaction force is zero, and it remains so if the 'overlap' term is non-positive. In contrast, the MD-type model (which we term the continuous interaction or CI) is clearly non-zero at $r=\sigma$. As is standard practice in MD, the model is truncated at some greater separation, r_c. A further consideration is the scaling of the model, *i.e.*, what value to assign to ϕ, the force at $r=\sigma$. We have chosen the gravitational attraction of the particle, *mg*. This results in the self-weight of one particle just resting vertically on top of another being exactly matched by the 'normal reaction' at equilibrium. The Hertzian model shares one feature of the CI in that it becomes progressively stiffer with increasing overlap. However, as has been shown for hopper flow and is further discussed herein for pouring, that Hertzian and spring models are fundamentally similar to each other and different from the CI. It emerges that the truncation point of an interaction is critical in its ability to reproduce certain experimental phenomena; principally, those which are related to the 'long-range connectivity' of granular media. Langston *et.al* [6] studied flow from hoppers using CI models in which the values of n, the interaction index, and r_c, the truncation separation, were varied. The formation of shearing or rupture zones during hopper discharge, an experimentally-verified phenomenon [8] in the simulations was found to be greatly enhanced by an extended range of interaction. We undertake a similar approach for heaps in this study, by comparing the results of CI pouring simulations with those performed with the linear spring model. We compare certain features of the poured heaps with the experimental and theoretical observations discussed above. For further details of the models employed and the relevant parameters within, see [7].

RESULTS

We compare the characteristics of poured heaps for simulations with the linear spring and CI models with experimental and theoretical observations. We are concerned with three related features of the heaps; the profile of internal contact stresses, the statistical distribution of the magnitude of these stresses and the distribution of normal forces on the supporting substrate.

Figures 1 and 2 show the network of normal contact forces within a poured heap for a CI and linear spring simulation respectively. In each case the heap has reached a near-stable state, containing around 1500 granules. The angles of repose (AOR) of the two heaps were very similar; 34° for the CI heap and 35° for the linear spring simulation. However, as can be seen, the internal force profiles (and other characteristics) of the two heaps are substantially different. A number of authors have suggested that the AOR may correlate with material properties such as the internal coefficient of friction. Alternately, authors such as Jenike [9] were highly skeptical of the AOR, suggesting that its popularity was more to do with the ease with which it could be measured rather than its usefulness. The results of this work tend to support the latter view.

In the CI simulation (Figure 1), the self-weight of the heap is largely propagated through the assembly along well-defined force chains. These form lines of principal stress which extend outwards from the upper regions of the free surfaces, through the heap to the base. There is clearly an uneven distribution of loads on grains, with preferential loading along the chains. Intersection of the lines of principal stress, in the region of the heap nearest the apex, results in the formation of arches. The self-weight is carried away from the central plane, resulting in rather low stresses on the grains towards the bottom centre of the heap. The lines of principal stress are inclined at an angle to the base greater than the AOR. This is in contrast with the earlier numerical investigations of Liffman *et.al* [10], who suggested that the principal stress lines were parallel to the free surface.

In contrast, for the linear spring simulation, the profile does not show arching; the stresses in the upper central region of the heap are rather low. In many cases, there are substantial components to the normal contact force in directions perpendicular to the principal stress direction as described above. This is particularly the case towards the lower central portion of the heap. Towards the base in this region, the load-bearing is more equally shared between supporting grains; there is little preferential loading.

We complement the analysis of the internal contact force profiles by considering the profiles of normal contact forces on the supporting base of the heap. This is a parameter which has been measured experimentally; for example, see [11]. The profiles for the two types of GD simulation are shown for comparison in Figure 3. As can be seen, the profiles for the two simulations differ markedly. This is to be expected, given the differences in the internal contact force profiles (Figures 1 and 2); the contact forces on the base are a reflection of how the self-weight is propagated through the heap, and hence of the internal contact force profile. Comparison with the work of Smid and Novosad [11] shows that, of the two profiles presented in Figure 3, the CI simulation reproduces experimental results much more closely. The qualitative agreement between the CI profiles and experiments is very good; profiles from both sources show a local minimum in the force below the apex of the heap, where the force is some 20-25% less than the maximum. Furthermore, profiles for the CI simulation as the heap grows (not shown here) also show good agreement with experiments, in that the dip is only a few percent of the maximum force at the beginning, increasing to over 20% as the heap grows.

As mentioned above, the forces on the base reflect the internal force profiles. The importance of uneven loading on grains within the heap has also been mentioned. Recent theoretical work on a model system by Edwards and Mounfield [12][13][14] suggests that self-weight stresses are propagated through the heap at an angle greater than the AOR if there is uneven sharing of loads on supporting grains. The model reduces to a mean-field approach, the result of which is that forces are transmitted along parallel chains, inclined at an angle β which is greater than the angle of repose α. The maximum force on the base occurs at the point where the longest possible line of 'principal stress' intersects the base, the point P. This is illustrated in Figure 4; only half of the model 'heap' is shown, because the profile is symmetrical about the central plane.

Given that two of α, β and the position of **P** are known, the model can be used to predict the third by simple trigonometry. It is straightforward to show that:

$$\frac{|OP|}{|OR|} = 1 - \tan\alpha \cot\beta \qquad (1)$$

For the continuous interaction simulation, the angle of repose α of the heap was 34°, and from visual inspection of Figure 1, the angle of propagation of forces β is approximately 60°. The model thus predicts that:

$$\frac{|OP|}{|OR|} = 1 - (0.674 * 0.577) = 0.611$$

The value of this parameter for the simulation can be found directly by visual inspection of Figure 3. The radius of the heap is approximately 40σ. The maximum force on the base is found approximately 25σ from the edge, and is thus located by: $|OP|/|OR| \approx 25/40 \approx 0.62$. The observed profile is in good agreement with the theoretical model in locating the maximum force on the base. The theoretical model is somewhat unrealistic in that it predicts zero normal load on the base below the apex of the heap, contradicting experimental results (and those of simulations). Nevertheless, the limited agreement with experiments suggests that the model may offer a partial explanation of the propagation of self-weight forces through heaps, and the simulation results support this view.

The profile of normal forces on the base for the linear spring simulation is markedly different, as shown in Figure 3. In light of the earlier discussion upon the internal contact force profile (Figure 2), self-weight loads are evidently shared more evenly between supporting grains than is the case for the CI. Consider a model system whereby loads are shared perfectly equally between supporting grains. In a mean-field sense the self-weight of the heap is propagated vertically downwards, since propagation in each lateral direction is equal and opposite. Thus, the force on the base at any point should reflect the vertical height of material directly above that point. The profile should therefore be triangular in shape, with a maximum below the apex. The profile for the linear spring simulation is similarly shaped, apart from a flattening near the centre. This is probably due to the heap being of slightly irregular structure near the apex, which is an inevitable consequence of the simulation approach. Thus, it appears that the linear spring model leads to near-equal sharing of loads throughout the heap, which is clearly at odds with theoretical and experimental observations.

DISCUSSION AND CONCLUDING REMARKS

We have shown via a series of simulation results, and comparisons thereof with experimental and theoretical studies, that realistic simulation of slow-shearing granular flows is better achieved using a continuous interaction model than a linear spring. We believe this to be due to the higher degree of long-range connectivity built into the CI model. Highly co-operative behaviour over lengthscales considerably greater than the single particle diameter is typical of slow-shearing flows of granular media. Nevertheless, our simulations show that building a greater degree of 'connectivity' into the interaction model <u>at the scale of the individual grains</u> leads to more realistic reproduction of granular behaviour. The flow fields within slow-shearing granular media are highly inhomogeneous. In certain regions there is negligible motion of grains relative to each other; locally, the material behaves like a solid body. Such regions have relatively high local density. In between such 'block gliding' zones are areas of high assembly deformation, high stress and low local density. Plots of both the internal contact force profiles (as in Figures 1 and 2) and density profiles (not shown herein) indicate that such heterogeneity is present in simulations with the CI, but the profiles are much more homogeneous for the linear spring simulations. Co-operative motion of large numbers of grains is lacking in the latter; particle motions are much more isolated and independent of each other.

The granule-granule interactions in these simulations are rather 'soft', i.e., particles are allowed to approach closer than a centre-to-centre separation of the nominal particle diameter σ, thus there is some overlap of the boundaries of adjacent granules. This could possibly be seen as a weakness of the model. However, work not included herein has led us to conclude that the stiffness of the interaction, reflected in the spring stiffness k or the interaction index n, is of secondary importance for the behaviour of the granular medium within the simulation. The truncation point of the interaction is far more significant. Thus, it may prove most appropriate to use a hybrid form of the continuous interaction model, whereby n assumes a high value for 'overlap' separations. This way, excessive overlaps are avoided, whilst the 'tail' of the interaction, giving long-range collective behaviour, remains.

The simulations described in this paper involve idealized systems in that they are two-dimensional, and the particle samples are monodisperse. Previous work on hoppers by Langston et.al [6] suggests that the principles of long-range connectivity discussed herein hold for three-dimensional slow-shearing flows also. The issue of polydispersity is somewhat difficult. Limited degrees of polydispersity, where the particle diameters vary by only a few percent, can be dealt with by assuming a nominal reference diameter for a pairwise contact as the arithmetic mean of the two particle diameters. However, if the size differences are greater than this, the resultant degrees of overlap are far too large to be considered realistic. To date, we have dealt with interactions of particles of widely differing size using the linear spring model. To some extent, there is a greater degree of connectivity throughout the assembly than is the case for the monodisperse linear spring simulations, due to the multiplicity of contacts between a large grain in the assembly and its smaller neighbours. Nevertheless, the behaviour of such assemblies would probably be better represented by a soft-sphere type model with an extended interaction range. Work continues in the development of such a model.

Long-range connectivity is a real phenomenon of slow-shearing granular flows. Explicit provision for long-range connectivity emerges as an essential aspect of the granular dynamics simulation strategy for such flows.

LITERATURE CITED

1. Bridgwater, J., Cooke, M.H. and Scott, A.M., "Inter-particle percolation: equipment development and mean percolation velocities", *Trans.Inst.Chem.Eng.*, **56**, 157 (1978)

2. Baxter, J., Tüzün, U., Burnell, J. and Heyes, D.M., "Granular dynamics simulations of two dimensional heap formation", *Phys.Rev.E.*, **55**, 3546 (1997)

3. Cundall, P.A. and Strack, O.D.L., "A discrete numerical model for granular assemblies", *Géotechnique*, **29**, 47 (1979)

4. Thornton, C. and Yin, K.K, "Impact of elastic spheres with and without adhesion", *Powder Tech.*, **65**, 153 (1991)

5. Ristow, G.H., "Simulating granular flow with molecular dynamics", *J.Phys.I France*, **2**, 649 (1992)

6. Langston, P.A., Tüzün, U. and Heyes, D.M, "Discrete element simulation of granular flow in 2D and 3D hoppers: dependence of discharge rate and wall stress on particle interactions", *Chem.Eng.Sci.*, **50**, 967 (1995)

7. Heyes, D.M., Tüzün, U. and Baxter, J, "Modelling of particle interaction laws in slow shearing granular flows", *AIChemE Symp.Ser.*, **93**, 113 (1997)

8. Bransby, P.L., Blair-Fish, P.M., James, R.G., "An investigation of the flow of granular materials", *Powder Tech.*, **8**, 197 (1973)

9. Jenike, A.W., "Storage and flow of solids", *Bulletin No.123 of the Utah Engineering Experiment Station* (1964)

10. Liffman, K., Chan, D.Y.C. and Hughes, B.D., "Force distribution in a two dimensional sandpile", *Powder Tech.*, **72**, 255 (1992)

11. Smid, J., Novosad, J., "Pressure beneath heaped bulk solids", *I.Chem.E.Symp.Ser.*, **63**, DV/V/1 (1981)

12. S.F.Edwards & C.C.Mounfield, "A theoretical model for the stress distribution in granular matter. 1. Basic equations", *Physica A*, **226**, 1 (1996)

13. S.F.Edwards & C.C.Mounfield, "A theoretical model for the stress distribution in granular matter. 2. Forces in pipes", *Physica A*, **226**, 12 (1996)

14. S.F.Edwards & C.C.Mounfield, "A theoretical model for the stress distribution in granular matter. 3. Forces in sandpiles", *Physica A*, **226**, 25 (1996)

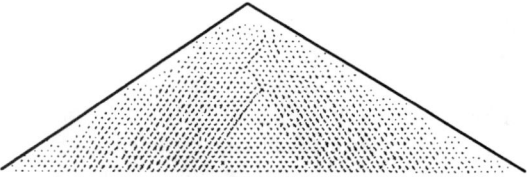

Figure 1: Distribution of normal contact forces for a continuous interaction simulation.

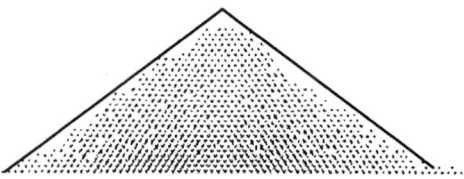

Figure 2: Distribution of normal contact forces for a linear spring simulation.

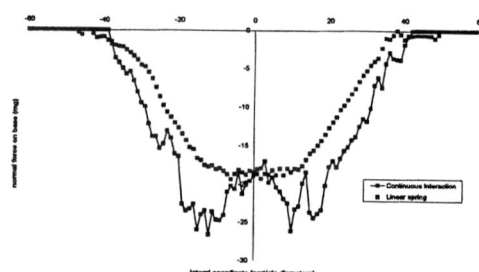

Figure 3: Profiles of normal contact forces on the base, for linear spring and CI simulations

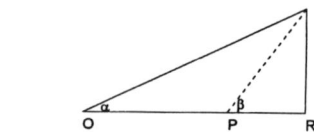

Figure 4: Geometry of model heap for theoretical analysis of stress propagation

Motion of Individual FCC Particles and Swarms in a Circulating Fluidized Bed Riser Analyzed via High-Speed Imaging

H. Hatano, H. Takeuchi*, S. Sakurai**, T. Masuyama** and K. Tsuchiya***
National Institute for Resource and Environment, 16-3 Onogawa, Tsukuba 305-8569, Japan
*Hokkaido National Institute for Research and Industry
**Tokai University
***The University of Tokushima

A circulating fluidized bed with a riser of 0.10-m inner diameter and 5.5-m height is used to study the flow structure of solid particles in the CFB riser. The solids circulation rate is fixed at 45.6 kg/m^2s; the superficial gas velocity is varied between 2.03 and 2.42 m/s to obtain an S-shaped and uniform distributions of particle concentration in the axial direction, respectively. Flow of fine particles (FCC; d_p=54μm) is visualized through a penetrative-type particle image scope (PPIS; 3 mm in diameter) with a high-speed video system. Visualization is performed in the fast fluidization and dilute transport regimes. All images are recorded at 4000 frames/s by dividing the full frame into four subframes. Dense particle swarms are observed in both the dense and dilute flow conditions. Typical particle swarms are analyzed to have thickness or diameter ranging from 0.1 mm to 2 mm and solids holdup exceeding at least 0.25 and possibly reaching that in a loosely packed condition. The velocities of individual particles and particle swarms are determined from the recorded images. While most of the individual particles have slower ascending velocities than the superficial gas velocity, the rising velocity of the particle swarms is found to be often higher than the superficial gas velocity. The frequency of the partcle swarms in the dense flow condition is higher than that in the dilute flow condition.

INTRODUCTION

Since the concept of circulating fluidization was introduced [1], particles in the CFB riser have often been analyzed by dividing the particle flow into a dispersed and swarming states. Particle swarms exhibit fluctuations in size, packed density, and velocity. Due to this time-varying complexity of the swarm structure, it is difficult to elucidate the role of the swarms in processes like coal combustion. The optical fiber probe technique has been applied to measure the velocity of swarm particles as well as that of dispersed ones (e.g., [2]). This technique, however, solely relies on a point probe that can only provide very local information, leading to less distinction between the two states of particle dispersion. It is also inadequate for observing transient behavior of the particle flow. An alternative technique, laser sheet lighting, has been applied to visualize the flow structure of a CFB. Horio and Kuroki [3] obtained images showing a three-dimensional network structure of U-shaped particle swarms. The solids concentrations examined in their study, however, were substantially lower compared to those under the normal CFB operating conditions.

In fast fluidization, the occurrence of ascending and descending particles in the riser has been well documented [4-8]; the flow structure caused by and the interaction between these particles, however, are still not clearly understood. It is very important to obtain detailed information on such particle behavior for modeling the heat and mass transfer, chemical reactions as well as hydrodynamics in the riser. Lim et al. [4] measured, from the outside of a riser, the descending velocity of particle clusters near the wall by video photography and compared the results with those based on a theoretical model. Takeuchi and Hirama [5] carried out a flow visualization study and identified the existence of dense particle swarms in the core region of a riser, although the solids concentration in the swarm was too high to obtain clear images. Hyre and Glicksman [6] used a sophisticated method to observe the fate of a single artificial cluster in the core region. Information, however, is still lacking to reveal the flow structure of the riser core with relatively high concentrations of particles.

The use of particle image scopes and a high-speed video

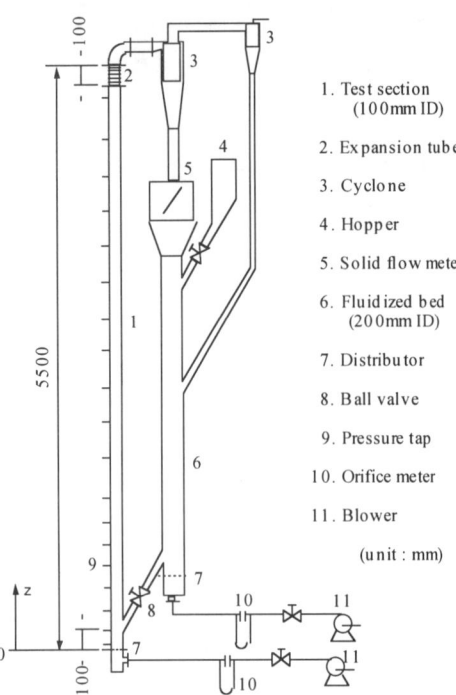

Fig. 1 Schematic of the circulating fluidized bed.

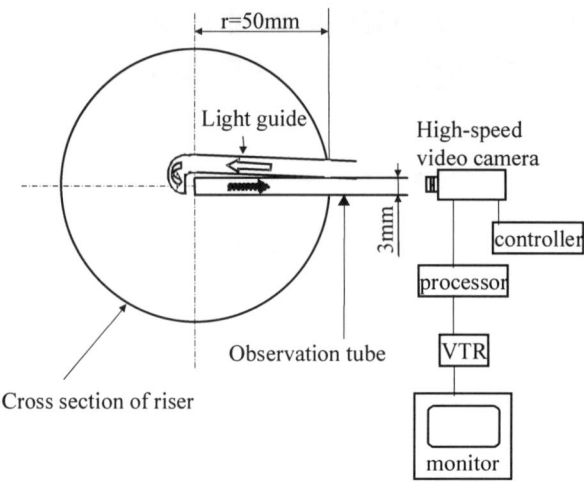

Fig. 2 Cross-sectional view of PPIS and visualization system.

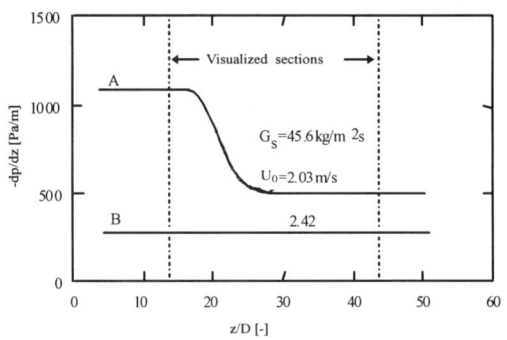

Fig. 3 Axial voidage distributions under two different flow conditions (dotted lines for axial locations of visualization).

system with image intensifier can enhance video images under relatively dark conditions, thus promising for the analysis of the local behavior/structure of particles including their shapes and velocities [7, 8]. The existence of parabolic/paraboloidal strands of particles in the core region was reported for a FCC particle system utilizing an internally-located view window [9]. In this study, the velocities of individual FCC particles and particle swarms are measured in two different flow conditions to obtain further information on the flow structure of fine particles in the core region.

EXPERIMENTAL

The CFB used for visualization experiments has a riser of 0.10-m ID and 5.5-m height as shown in Figure 1. FCC particles are used as the circulating solids; the mean diameter is 0.057 mm and the bulk density is 930 kg/m^3. Figure 2 shows the visualization system involving a penetrative-type particle image scope (PPIS) [7]. The PPIS is inserted in the core region horizontally. The image guide of the PPIS is of 3 mm in OD and is connected to a high-speed video camera (Kodak Ektapro EM1012). The resolution of the camera is 238(H) × 192(V) pixels. The PPIS images are recorded mostly at 4000 subframes/s with an exposure time of 0.05 ms. Images are stored on the memory of the video processor over a total recording time of 1.6 s. The stored images are then transferred to a video cassette recorder at a rate of 30 frames/s. Xenon light is introduced from the outside of the riser through an optical fiber light-emitting device.

The visualization point/zone is defined as a space between the tip of the light-emitting device and the tip of the image fiber(5 mm gap). The visual observation is made at two elevations, z/D = 14.5 and 45, and at four radial locations, r/R = 0, 0.2, 0.4, 0.6 and 0.8. At each visualization point, 1000 frames out of the video recording over 1.6 s are used for the analysis. Due to limited quality of images which is not sufficient to clearly identify the edges of particle swarms and single particles, the frame analysis heavily relies on careful recognition of these edges by naked eyes.

Experiments are performed for a fixed value of the solids circulation rate, G_S = 45.6 kg/m²s, at two different superficial gas velocities, U_0 = 2.03 and 2.42 m/s. The lower gas velocity is in fast fluidization (Condition A), the higher gas velocity in dilute transport (Condition B). Two conditions can be represented by the corresponding voidage profiles along the axial direction of the riser; as shown in Figure 3, the former condition is characterized by an S-shaped profile, the latter by a uniform profile. The axial voidage profile is determined from the measurement of the axial pressure gradient (see Figure 1).

RESULTS AND DISCUSSION

Figure 4 shows typical consecutive images, recorded at 200 frames/s, of FCC particles flowing in the central region of the riser under Condition A. Over a period of 0.2 s, the solids flow structure changes drastically from one extreme phase, uniform dispersion, to the other, dense swarming. Omitted in the figure between 5.985 s and 6.085 s, the images in the whole view appear totally dark, signifying the passage of a large, dense particle swarm. After its passage, small particle swarms of thickness less than 1 mm can be identified. Here the thickness (or diameter) of a particle swarm is defined as the vertical

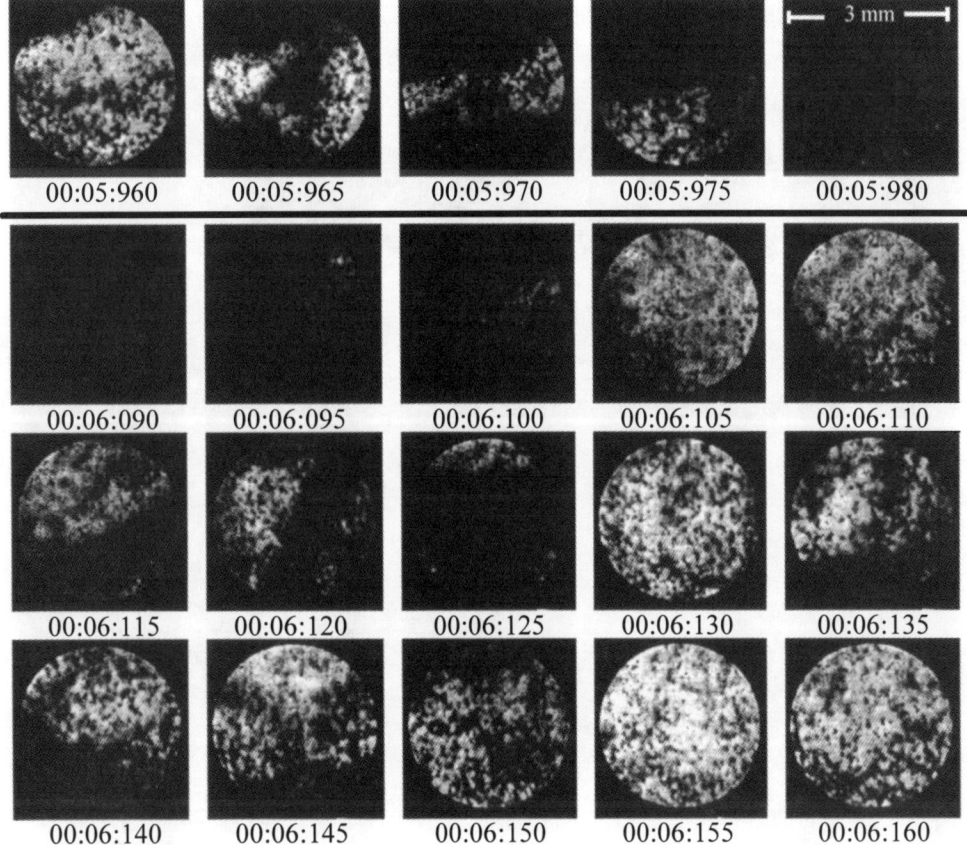

Fig. 4 Consecutive images in the riser core region (dense condition).

dimension of the swarm; its lateral dimension may exceed the size of the present viewing window, 3.0 mm (usually on the order of 10–20 mm [9]).

Images of small particle groups, isolated in the view, are shown in Figure 5. The analysis of these images indicates that the shape of these particle groups is generally characterized by parabolic strands, and the solids holdup inside the group could reach that in a fixed bed. Takeuchi et al. [9] observed the typical size of parabolic strands to be a few mm in thickness using a 22-mm viewing window. Although the measured size of the particle strands may be subjected to the window size, it is possible that the smaller strands observed in this study are part of a large strand of three-dimensional network structure.

Figure 6 shows the time variation of the particle ascending velocity measured along the central axis in the upper section under Condition B. For individual particles, the velocity is confined below the superficial gas velocity (Figure 6a). For particle swarms, the data are largely scattered and the velocity often exceeds the superficial gas velocity, sometimes as high as 2 times of U_0 (Figure 6b). This trend of having higher ascending velocities of particle swarms was also observed in the previous study [9].

Some individual particles have relatively large values of the lateral velocity component as shown in Figure 7a, although many others have negligible values or only the vertical component. An average value of the absolute lateral velocities given in Figure 7a is 0.15 m/s. It is also realized that particle swarms have the lateral velocity component (Figure 7b); the number of data points, however, is limited due to difficulty in measuring it for swarms larger than the view area or ascending very fast. The data represented by small filled circles in Figure 7b

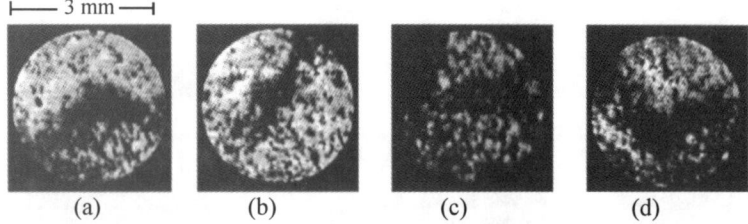

Fig. 5 Images of small particle swarms (dense condition).

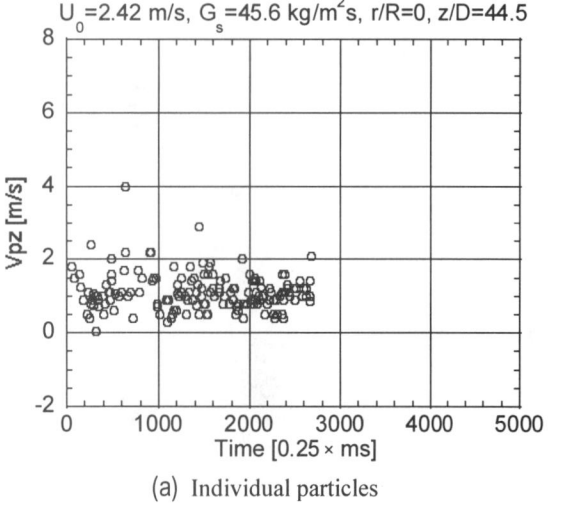

(a) Individual particles

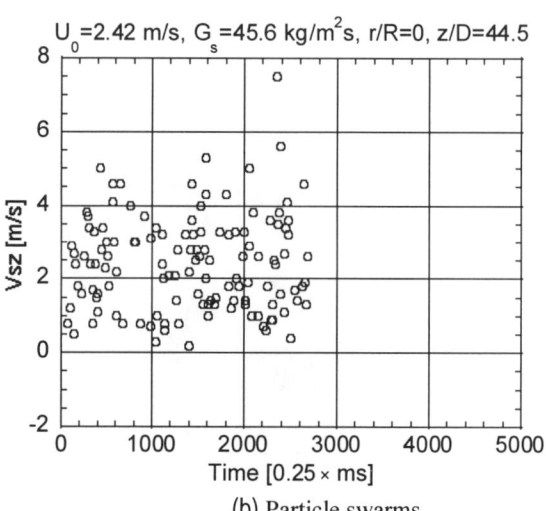

(b) Particle swarms

Fig. 6 Time variation of particle velocity (dilute condition).

reflect this limitation (i.e., no measurement of V_{sx}). Similar trends in the particle motion are obtained for particles flowing in the lower section.

Under the dense flow condition, particle swarms which are more densely packed pass the measuring section more frequently, especially in the lower section of the riser. This imposes additional difficulty in obtaining reliable data. Nevertheless, the motions of individual particles and particle swarms can be traced by the careful frame-by-frame analysis. The observed velocity distributions are quite similar to those under the dilute flow condition. Individual particles in the dispersed phase are closely associated with the gas flow, while some of the particle swarms, especially larger ones, flow independently from the gas flow.

Figure 8 shows the relationship between the ascending velocity of particle swarms and the vertical dimension, or thickness, of the swarms. Measured in the upper section, the velocity exhibits a positive dependence on the thickness (Figure 8a); such a relationship can hardly be found in the lower section (Figure 8b). This finding, however, may not be conclusive since only one dimension is measured and used to represent the size of the particle swarms. For the given data sets, the measured maximum thickness is less than 2 mm due to the limited diameter of the viewing area of the PPIS. With a large viewing area, larger swarms (parabolic strands) have been observed [9]. In addition, the particle swarms in the riser have three-dimensional network structures, characterized by a variety of geometry

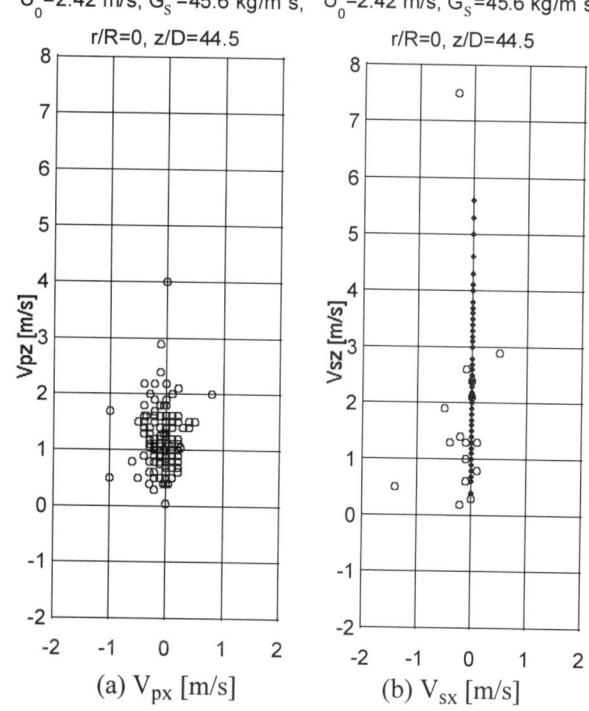

Fig. 7 Particle velocity components for (a) individual particles and (b) particle swarms (dilute condition).

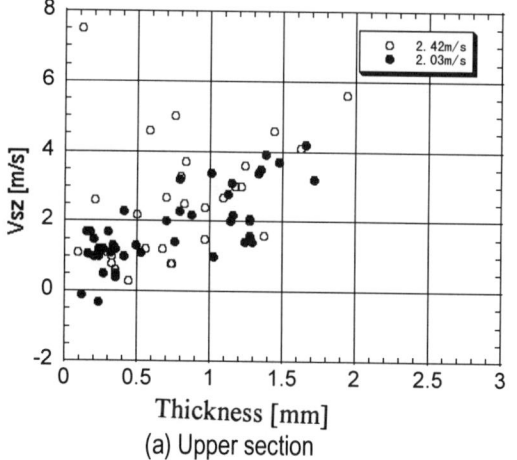

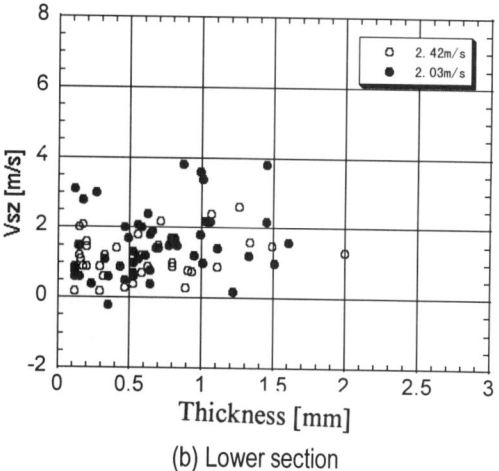

Fig. 8 Relationship between the velocity and size of particle swarms observed in (a) upper and (b) lower sections (open symbols: dilute condition; filled symbols: dense condition).

(strands, clusters, dense packets, etc), warranting further research.

CONCLUSIONS

While most of the individual particles have slower ascending velocities than the superficial gas velocity, the rising velocity of the particle swarms often exceeds the superficial gas velocity. The frequency of the particle swarms in the dense flow condition is higher than that in the dilute flow condition. The solids holdup of the particle swarms exceeds at least 0.25 and possibly reaches that in a loosely packed condition.

NOTATION

D	riser diameter	m
d_p	particle diameter	m
G_S	solids circulation rate	kg/m²s
p	pressure	Pa
R	riser radius	m
r	radial coordinate	m
U_0	superficial gas velocity	m/s
V_{px}	lateral velocity of individual particles	m/s
V_{pz}	vertical velocity of individual particles	m/s
V_{sx}	lateral velocity of particle swarms	m/s
V_{sz}	vertical velocity of particle swarms	m/s
z	vertical coordinate above distributor	m

LITERATURE CITED

1. Yerushalmi, J., and N. T. Cankurt, "High Velocity Fluidization", *CHEMTECH*, **8**, 564 (1978).

2. Horio, M., H. Ishii, and M. Nishimuro, "On the Nature of Turbulent and Fast Fluidized Beds", *Powder Technol.*, **76**, 229 (1992).

3. Horio, M., and H. Kuroki, "Three-Dimensional Flow Visualization of Dilutely Dispersed Solids in Bubbling and Circulating Fluidized Beds", *Chem. Eng. Sci.*, **49**, 2413 (1994).

4. Lim, K. S., J. Zhou, C. Finley, J. R. Grace, C. J. Lim, and C. M. H. Brereton, "Cluster Descending Velocity at the Wall of Circulating Fluidized Bed Riser", *Circulating Fluidized Bed Technology V*, p. 218 (1996).

5. Takeuchi, H., and T. Hirama, "Flow Visualization in the Riser of a Circulating Fluidized Bed", *Circulating Fluidized Bed Technology III*, p. 177 (1990).

6. Hyre, M., and L. R. Glicksman, "Measurement of the Breakup of Clusters in the Core of a Circulating Fluidized Bed", *Circulating Fluidized Bed Technology V*, p.224 (1996).

7. Hatano, H., and N. Kido, "Microscope Visualization of Solid Particles in Circulating Fluidized Beds", *Powder Technol.*, **78**, 115 (1994).

8. Hatano, H., S. Matsuda, H. Takeuchi, A. T. Pyatenko, and K. Tsuchiya, "Local Interactive Patterns of Dispersed and Swarm Particles in a Circulating Fluidized-Bed Riser", *Ind. Eng. Chem. Res.*, **35**, 4360 (1996).

9. Takeuchi, H., A. T. Pyatenko, and H. Hatano, "Flowing Behavior of Particles in the Riser of a Circulating Fluidized Bed", *Circulating Fluidized Bed Technology V*, p.164 (1996).

Mechanism of Solid Flow in a Closed Loop Circulating Fluidized Bed With Secondary Air Injection

J.H. Kim, K. Shakourzadeh and J.F. Large
Département de Génie Chimique, Université de Technologie de Compiegne
BP 649, 60206 Compiegne, France.

The aim of this work is to study the mechanism of solid circulation in a CFB pilot as a function secondary air flow rate. A rectangular column of 7 m height equipped with a U type siphon is used for this purpose. Circulating solid flow rate and pressure drop, are measured under a set of different combinations of primary air flow rate, secondary air flow rates and solid inventory.

The results obtained showed that the solid circulating phenomenon depends on different limiting steps like feeding step (dense bed), siphon circulating capacity and suspension saturation capacity.

INTRODUCTION

Actually, there are very few existing theories concerning the solid flow rate in closed loop CFB. With « Closed Loop» we mean that solid circulating rate is free and can be only cut for flow rate measurements. While in an « Open Loop » configuration, an intermediate hopper is used to control the solid feed rate. CFB reactors used in coal combustion processes are generally of « Closed Loop » type.

These reactors has always a secondary air feed, introduced some meters above the fluidization grid, in order to control the combustion quality and reduce Nox production. However, the division of total air flow to primary and secondary air, affects the flow structure at the lower level of the reactor. Few works have been done to study the effect of secondary air on the flow structure [1,2,3].

In a previous work [4], we discussed about the hydrodynamic behaviour of open loop circulating fluidized beds with secondary air injection. We concluded that in this case the rate of secondary air injection (compared to total air) modifies slightly the flow pattern at the stabilised region (above 3 m height) of the CFB.

In the present work however, we show that for « Closed Loop » CFB the distribution of air feed at primary or secondary position affects strongly the flow properties and the solid circulating flow.

EXPERIMENTAL INSTALLATION

Figure 1 shows the CFB unit used in this work. The pilot unit has been described in details in [4]. We recall here some general aspects of the set-up. The column has a rectangular 0.286 x 0.176 m cross section and is 7m high. The solid flow rate is measured separately in a solid return line. The column is equipped with Plexiglas windows and some 36 pressure taps. Gas and solids are separated by a primary charged cyclone, a secondary standard cyclone and a set of filters. Solid particles are fed through a standpipe with a siphon for controlling the flow rate. In addition to the geometrical similarity between the pilot and CFB combustors, we use cracking catalyst (mean particle size of 68 microns) at ambient temperature, to operate at similar Archimede numbers and similar solid concentrations. The fluidization air can be distributed between the primary or the secondary air injections. The industrial operating conditions corresponds to 2/3 of total air to the secondary and only 1/3 to the primary air injection.

A siphon is used in the return line in order to establish the pressure drop between the bottom of the fluidized bed and the air exit at cyclone level. The siphon is found to be a limiting step for high fluidization velocities.

Solid circulating flow rate is measured by means of an external weighted hopper. Note that this measurement technique needs the external line to be cut. However, the variation of hopper weight remains linear during 10 to 15 seconds (before the column concentration begins to decrease), where the slope of changing weight gives the solid flow rate. The measurement error is found to be less than 10%. The only problem with this technique is that after each measurement the pilot needs some 15 minutes to regain the stationary operating conditions.

Local solid concentration and particle velocities are measured by means of « Optical Fibres »at three different levels (P0, P1 and P2 on figure 1). These measurements give useful information about the structure of the solid flow and let us to develop some our hypothesis presented at this work. However, to avoid to be too long, these results are not discussed here and will be presented in future.

RESULTS AND DISCUSION

Among different variables of the system, the gas velocity 'U_g' (total air flow / column cross section), secondary air / total air ration 'SA/TA' and solid

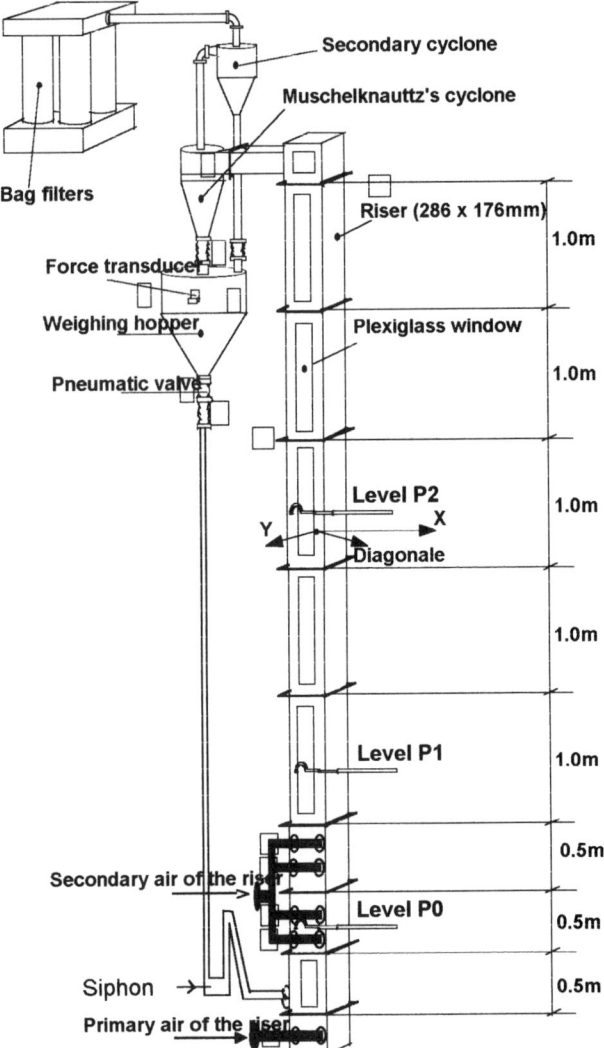

Figure. 1 The circulating fluidized bed pilot.

inventory 'M' are chosen as initial variables and the effect of their variation on the solid circulating rate and on the solid concentration in the column is studied. Solid local and mean velocities are measured as well, in the column. However, we will limit here our discussion on the first three variables.

Figure 2 shows the variation of solid circulating flow rate and solid mean concentration in the column as function of air velocity in the column. From this figure, it can be seen that for low gas velocities, both variables increase linearly, then there is a stagnation zone (circulating rate decreases even slightly) then they

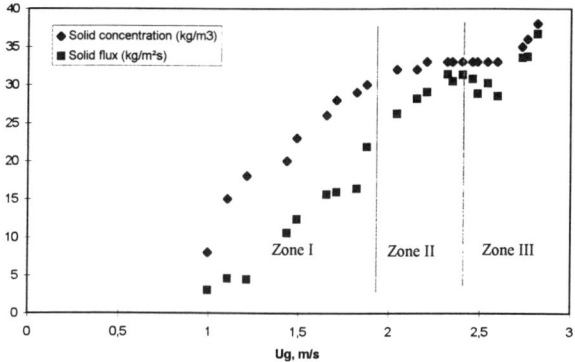

Figure 2 - Variation of solid circulating flow rate and solid mean concentration as a function of gas flow rate, with M= 35 kg and SA/TA=0.

increase again. The solid flow seems to be regular at the first zone but fluctuating and irregular at the second. Some irregular behaviour of siphon have been noted as well at this second zone (probably because of pressure fluctuations in the column). At higher velocities the flow structure changes again and it becomes regular again. This third zone, however, could not be fully studied because of gas velocity limitation.

Similar results are obtained at lower solid inventory. These results are shown on the figure 2, where we note that the circulating flow rate is higher at higher solid inventories. Note that the global form of the curve is independent of the solid inventory and similar zones can be defined again on these curve. The transition velocities are the same in all cases (2.3 m/s and 3 m/s respectively between zones I- zone II and zone II-zone III.

As we mentioned at the beginning of this paper, the main goal of this work is to define the effect of secondary air on the circulating flow rate. Figure-4 shows the variation of solid circulating flow rate as a function of gas velocity at different SA/TA ratios.

Figure 4 is an important diagram that is obtained from different series of measurements. Note that the solid flow rate is null if all the air is injected into the secondary level (SA/TA=100%) and values at SA/TA=0 are a part of results already shown on figure 2.

It can be seen from this diagram that putting air to the secondary level (rather than primary level) causes a decrease in the circulating flow rate. That means, that the main controlling step of solid transport in the column is at the dense to dilute be transport.

At very low velocities, a slight distribution of air at secondary level, seems to improve the circulating rate, probably because it prevents the some clusters ejected from dense to dilute bed to fall down.

Finally to verify the role of secondary air at constant primary air feed (that means that the lower dense bed feeds the upper dilute bed by a constant rate of solid clusters), solid flow and solid concentration are measured as functions of additional secondary air.

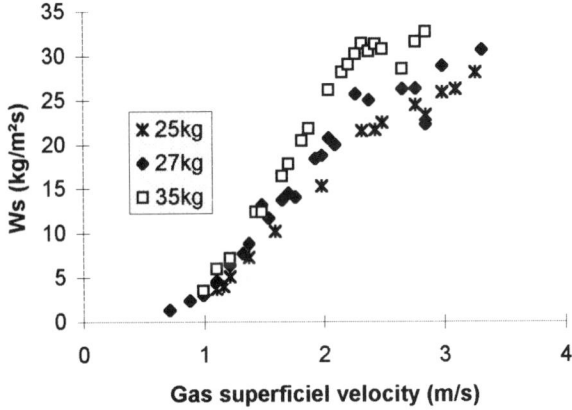

Figure 3 - Effect of solid inventory on the circulating flow rate

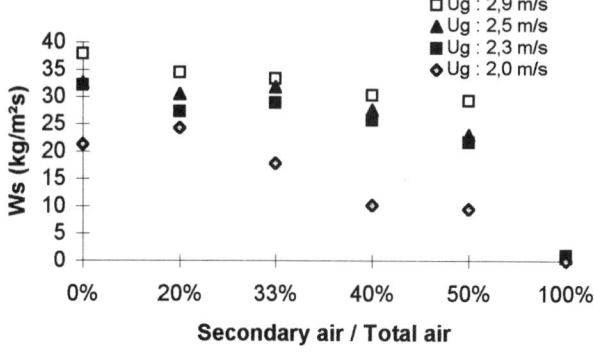

Figure 4- Effect of secondary air on solid flow rate (M=35 kg).

These results are shown in figures 5 and 6. The results on figure 5 show that solid flow rate increases and reaches to a limit value as the secondary air increases (this is independent from the primary air flow rate varying from 1.5 to 2.5 m/s). Solid concentration however, increases first, with the rise in secondary air, then decreases slightly for higher values of secondary air.

The analysis of above results lets us consider that the lower dense bed behaves as a solid feeder to the upper dilute bed. As some solid clusters fall down (especially near the wall), a secondary air feed prevents some clusters to fall back into the dense bed and improves gradually the solid transfer between two parts. Nevertheless the final value of solid flow rate cannot overpass the upward solid flow produced by the lower bed. Again for solid concentration, it increases in the first part of the diagram (figure 5) then decreases because the solid flow rate becomes constant but solid mean velocity continue to increase because of increasing gas velocity (we remember that the secondary air is fed in addition to the primary air so that the total air flow increases).

CONCLUSION

The present work tends to explain the mechanism of solid transport through closed loop Circulating Fluidized Beds.

In the light of results obtained, we can conclude that the lower part of the bed behaves as a feeder to the upper part. Meanwhile, the upper part of the bed (the dilute bed) has a limited capacity in solid concentration and discharges a part of the solid fed by the lower part (Figure 7). The solid flow rate is firstly controlled by the primary air that imposes the main solid (upward) flow F^+. The secondary part of air can only affect the downward solid flow F^- that decreases with a rise in secondary air.

The definition of a an empirical correlation, giving solid flow rate as a function of other variables doesn't seem to be useful for these complicated systems. This rule must be defined by writing a complete set of force balance equations that will be the subject of a future publication.

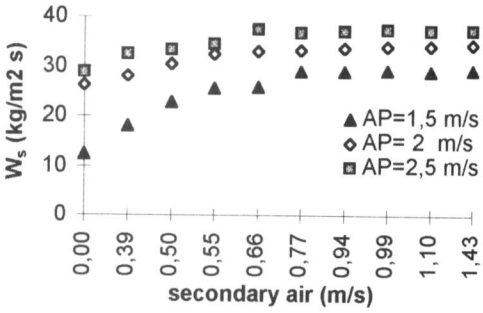

Figure 5 - Solid flow rate as a function of secondary air at different primary air flow rate.

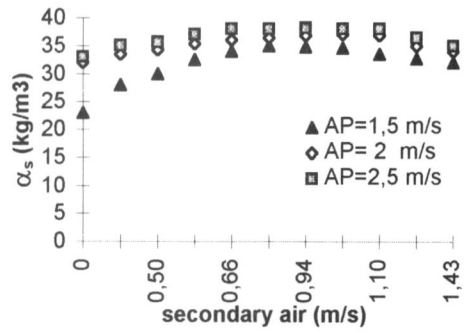

Figure 6 - Solid mean concentration in column as a function of secondary air flow and at different primary air flow rate.

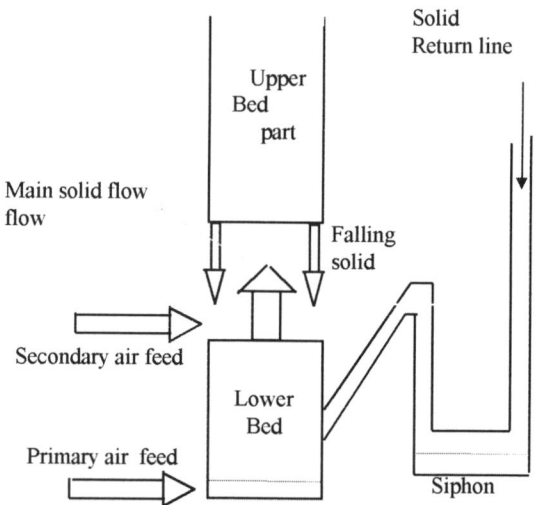

Figure 7 - Mechanism of solid transport in a closed loop CFB with secondary air feed.

REFERENCES

1. Wang X.S. and Gibbs B.M., « Hydrodynamics of CFB with secondary air injection », CFB Tech. III, Ed. Basu P., Horio M., and Hasatani M., Pergamon Press, Oxford, p225 (1991)

2. Arena U., Camarota A., Mazocchella A. & Massimilla L., « Hydrodynamics of a circulating fluidized bed with secondary air injection »; Fluid.Bed.Combust.., vol.2 (1993), p899-905

3. Cho Y., Namkung W., Kim S. D. & Park S., « Effect of secondary air injection on axial solid hold-up distribution in a circulating fluidized bed », J. of Chem. Eng. of Jap., vol.27, no.2 (1994), p158-164

4. Aguillon J., Shakourzadeh K., Guigon P., Powder Tech., 83 (1995) p.79-84

5. J. Militzer, K. Shakourzadeh, J.P. Hebb & G. Jollimore, « Solid particle velocity measurement », Proc. of Fluidisation VII, Poter & Nicklin ed., Brisbane -Australia, May 3-8 (1992), p.763-769

LIST OF SYMBOLS:

Symbol	Description	Units
F^+	Upward solid flow	$kg/s.m_2$
F^-	Downward solid flow	$kg/s.m_2$
M	Solid inventory in the column	kg
PA	Primary air flow rate	m/s
SA	Secondary air flow rate	m/s
TA	Total air flow rate	m/s
U_g	Mean gas velocity	m/s
W_s	Solid flow rate	$kg/s.m^2$

Non-Intrusive Solids Velocity Measurement In a Downcomer

Stephen Tallon and Clive E. Davies
Industrial Research Limited, P.O. Box 31-310, Lower Hutt, New Zealand

Bernard Barry
Institute of Geological and Nuclear Sciences, Lower Hutt, New Zealand

The velocity of solids flowing in a downcomer can be measured on-line and non-intrusively using sound waves. The attenuation of sound through a bed of particles is sensitive to the packing structure of the material and can be used to detect the passage of small local variations in the voidage. This forms the basis of a method used to measure solids velocities, and results are given for wheat, plastic pellets, and sand. These velocities are compared to independent mass flow rate measurements, and velocity profiles measured using radioactive tracer particles.

In many chemical industries involving the use of particulate materials the particles flow under gravity through vertical conduits, either in transportation, or as the feed to a processing vessel such as in circulating fluidised beds. Knowledge of the flow rate of the material is often useful for controlling the process and for maintaining product inventories, but is a variable which has eluded successful measurement with a method that is both on-line and accurate. Several methods have been described for measuring the flow rate (1)(2)(3)(4)(5) but have a number of draw backs including limited accuracy, mechanical complexity, or the need to introduce an object into the flow stream. This paper describes a method for measuring the velocity of the solids using acoustic measurements.

Sound waves are well suited to analysis of particulate systems. They are inexpensive to generate, the transducers do not need to intrude into the particle flow stream, and the velocity and attenuation of the waves is sensitive to particle properties including the size, shape, and voidage (6)(7). The method described here for measuring the solids velocity uses measurements of the intensity of sound waves passing across the diameter of a vertical downcomer. The intensity of the waves is modulated by small variations in the packing structure of the material as it flows past the measurement point. By measuring these variations in sound intensity at two axially spaced points in the column, the passage of characteristic patterns in the packing structure can be detected. Cross correlation (8) can be used to calculate the time taken for the packing structures to pass between the two measurement points. A characteristic velocity for the particles associated with these packing structures can then be calculated from the distance between the points and the transit time. Results are presented for two common industrial Geldart group D particles, wheat and plastic pellets, and for a typical group B material, sand.

When particulate materials flow through a downcomer there is slip between the particles and the pipe wall which can result in a shear zone near the wall where the particles travel at a lower velocity than in the centre of the pipe. This has been reported by many researchers, for example (9)(10)(11)(12). The shear zone is reported as typically around five particle diameters wide, and having a velocity between 5% and 50% lower at the wall than in the center. The effect of particle size and wall roughness on the velocity profile is discussed by Nedderman and Laohakul (9) and the effect of particle frictional parameters is discussed by Takahashi and Yanai (10). The behaviour of some materials is however hard to predict and we consider other factors such as particle shape to be significant as well. In the work here, velocity profiles across the

---

Industrial Research Limited, Gracefield, Lower Hutt, New Zealand

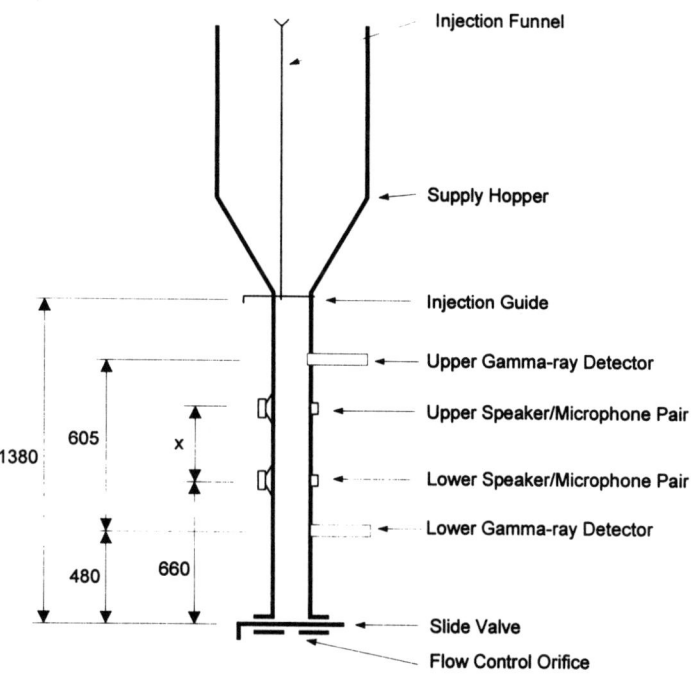

Figure 1 - Experimental Apparatus. Dimensions in mm

downcomer are measured for the two Geldart D materials, wheat and plastic, using radioactive tracers injected at different radial positions. The two materials have visibly different flow properties as well as different velocity profiles, and the results are discussed. The effects of these different flow properties on the acoustic velocity measurement method are also discussed.

EXPERIMENTAL

The apparatus used is shown in Figure 1. The perspex column had an internal diameter of 150mm for runs with the wheat and plastic pellets. Acoustic measurements of the velocity of sand were made in a 63mm ID column. Short 20 cycle pulses of 5000 Hz sine waves were generated using piezoelectric speakers at a rate of 20 pulses per second and with a 4ms lag between the top and bottom speaker to avoid interference. The sound intensity for each pulse was calculated from the Root Mean Square (RMS) value of the digitised signal from the microphones, and the solids velocity was then calculated by cross correlating the intensities at the two points. The distance, x, between the top and bottom measurement was either 100mm or 300mm for the larger particles. For experiments with sand the microphones were 10mm apart and the acoustic signals were generated using only the lower of the two speakers.

To measure the velocity profile, small radioactive gold particles were sealed into a few individual wheat and plastic particles. The tagged particles were then dropped through an injection funnel, the outlet of which was placed in different radial positions. Their passage through the system was detected at the two points indicated in Figure 1 using NaI(Tl) scintillation gamma-ray detectors. The detectors were shielded to give a narrow field of view. The particle velocity at each radial position was then calculated from the spacing between the detectors and the times at which the tagged particles passed the detectors.

The flowrate through the downcomer was controlled using interchangeable orifices. For each orifice and material the total flow of solids was recorded over a known time to get a mass flow rate. The density of the solids was also recorded in-situ from the mass of material contained in a measured volume of the pipe by stopping the flow when the hopper had drained and the solids formed a free surface in the downcomer. A mean solids velocity was calculated

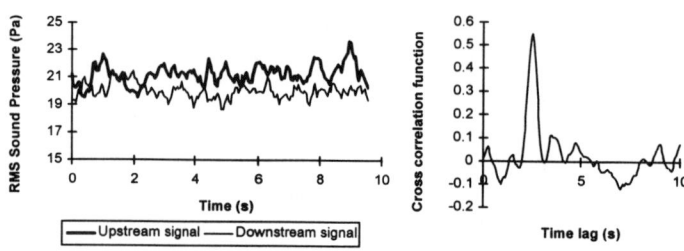

Figure 2 - Typical acoustic measurements and cross correlation for wheat in 150mm diameter downcomer. 100mm separation between signals.

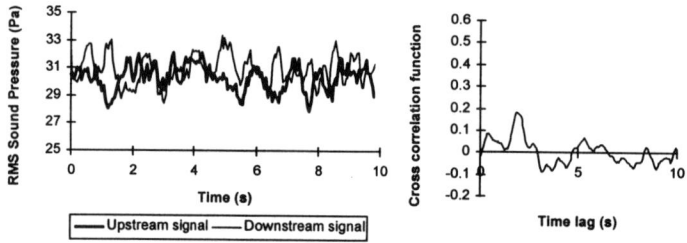

Figure 3 - Typical acoustic measurements and cross correlation for plastic pellets in 150mm diameter downcomer. 100 mm separation between signals.

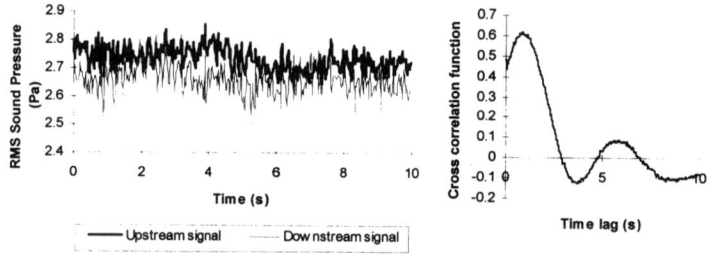

Figure 4 - Typical acoustic measurements and cross correlation for sand in 63mm diameter downcomer. 10 mm separation between signals.

from these values for comparison with the other velocity measurements.

The wheat particles typically had a length of about 7 mm and diameter 3 mm. The plastic pellets were short cylinders typically 3.5 mm in diameter and 2 mm in height. The sand had a mass mean diameter of about 150 μm.

RESULTS AND DISCUSSION

Figures 2, 3 and 4 show typical recordings of the RMS sound pressure level and the corresponding cross correlation functions for, respectively, wheat, plastic pellets, and sand. Fluctuations of greater than 10% of the mean sound pressure are apparent for all the materials, indicating the sensitivity of this measurement to very small changes in the packing properties of the materials. Spectral analysis of the signals shows a broad band of frequencies for all the materials and no clear periodicity. Cross correlation of the signals however shows strong correlation between the two points, as indicated in the figures. Wheat gave the strongest correlation, Figure 2, even over a 300mm distance between the measurement points. Qualitatively the plastic pellets flowed differently to the wheat. Visual observation of the flow through the column wall suggested the wheat grains were locked together in plug flow even at the wall, whereas the plastic pellets were free to move relative to each other as they flowed through the pipe. The effect of this relative movement on the coherence of the small voidage variations, is indicated by the weaker correlation of the signals for plastic, Figure 3.

The sand also shows some relative motion between the particles and because the particle size is smaller, the signals could only be correlated over a 10 mm distance. In this case the width of the acoustic path between the speaker and the microphone is significant, as evidenced by the broad correlation peak in Figure 4. The resolution of the velocity is consequently low, and the results presented here for sand are limited to quite low solids velocities.

The velocity profiles for wheat and plastic pellets, measured using tagged particles, are plotted in dimensionless form in Figure 5. The velocity has been normalised by a mean velocity over the flow area calculated from the velocity measurements at each radial position. There is a distinct difference between the two materials. The profile for wheat is flat, indicating plug flow right up to the pipe wall, as could be anticipated for elongated particles where the solids are locked together. The plastic by contrast showed a sharp decline in velocity at the wall, and a lot more scatter. The shear zone is nominally five particle diameters wide, consistent with the values commonly reported in the literature. In the center of the column the profile is relatively flat and the velocity is about 6% higher than the mean.

The velocities calculated using the acoustic method are shown in Figures 6, 7, and 8 along with the velocity at the center line measured using the tagged particles, and a velocity measured by visually observing a single particle travelling down the wall of the downcomer. These velocities are plotted against a mean velocity, estimated by dividing the mass flow rate by the measured density and the area of the downcomer. The results for wheat, Figure 6, all agree to within 2%. The velocity at the pipe wall is the same as that measured at the center of the pipe and the acoustic method gives an accurate measurement. For plastic pellets, Figure 7, the wall velocity is about 25% lower than the mean value and the center line velocity is about 6% higher. These results are consistent with the velocity profile shown in Figure 5. The acoustic method gives velocities consistent with the velocity of

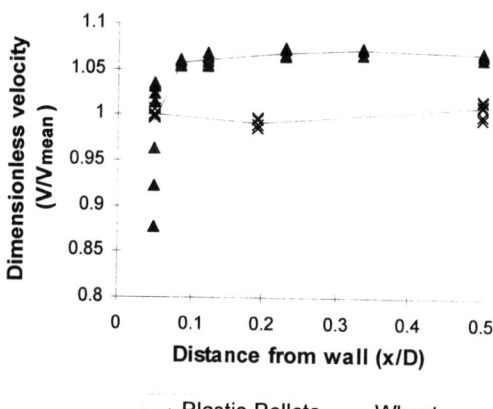

Figure 5 - Dimensionless velocity profiles for wheat and plastic pellets

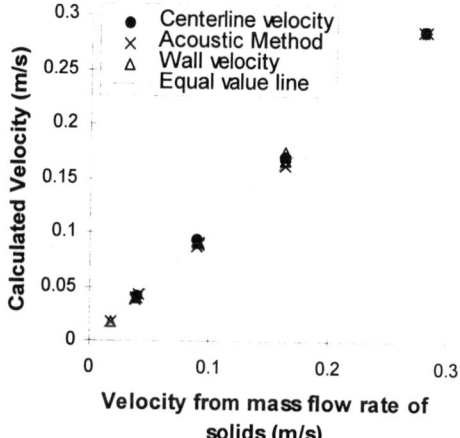

Figure 6 - Velocity measurements for wheat

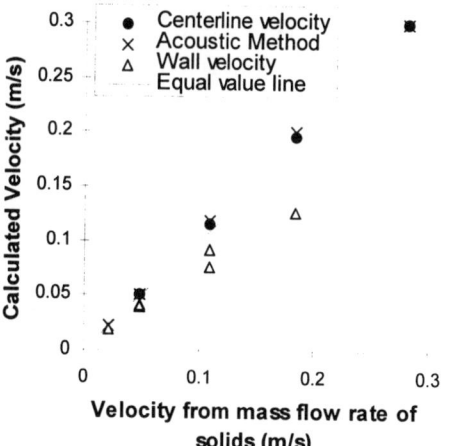

Figure 7 - Velocity measurements for plastic pellets

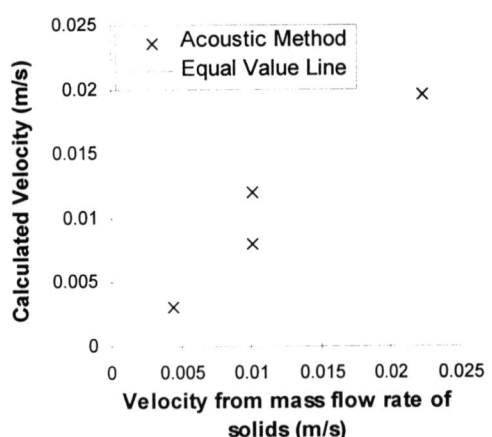

Figure 8 - Velocity measurements for sand

the plug flow region in the center of the column. For wheat this is the same as the mean velocity over the whole column, but where a shear zone at the wall exists the acoustic measurement returns a slightly higher value. Variations in the local particle packing that are detected in the central plug flow region of the column are expected to give stronger correlation than those in the shear zone near the wall where there is a greater degree of shear and distortion of the particle packing structures. Consequently the particle velocity measured using the acoustic technique has a stronger correlation with the central plug flow velocity than the mean velocity calculated from the mass flow rate of solids.

The effect of the particle size, and of the axial spacing between the sound measurements, on the solids velocity measurement is discussed by Tallon and Davies (13).

CONCLUSION

A method for measuring the velocity of solids in a vertical downcomer has been demonstrated using measurements of fluctuations in the intensity of sound waves passing through the solids. The method is accurate to within 2% of independently calculated values of the plug flow velocity of the solids. For materials such as plastic pellets there is a shear zone near the wall where particle velocities are lower than through the rest of the column. Measurements of the velocity profiles of the different materials have confirmed that the acoustic velocity measurement gives the velocity of the central region rather than a mean velocity. Consequently, for plastic pellets, the acoustic measurement returns a value about 6% higher than the mean value over the whole cross section. Qualitative differences in the flow of the different shaped particles have also been noted.

LITERATURE CITED

1. **Burkell, J.J., Grace, J.R., Zhao, J. And C.J. Lim,** Circulating Fluidized Bed Technology II, Proceedings of the Second International Conference on Circulating Fluidized Beds, Compiegne, France, edited by Basu, P., and J.F. Large, Pergamon Press, Oxford, pp 501-509, (1988).

2. **Davies, C.E., and B. J. Harris,** Fluidization VII, Proceedings of the Seventh Engineering Foundation Conference on Fluidization, Brisbane, Australia, edited by Potter, O.E., and D.J. Nicklin, pp 741-747, (1992).

3. **Judd, M.R., and H.W. Bernhardt,** Circulating Fluidized Bed Technology III, Proceedings of the Third International Conference on Circulating Fluidized Beds, Nagoya, Japan, edited by Basu, P., Horio, P., and M. Hasatami, Pergamon Press, Oxford, pp 621-626, (1991).

4. **Patience, G.S., Chaouki, J., and B.P.A. Grandjean,** Powder Technology, **61**, pp 95-99, (1990).

5. **Liu, J., and Bowen Huan,** Powder Technology, **82**, 2, pp 145-152, (Feb. 1995).

6. **Tallon, S., and C.E. Davies,** "Attenuation in beds of particulate materials and application to measurement of flow rate", Fluidization IX, Proceedings of the Ninth Engineering Foundation Conference on Fluidization, Durango, Colorado, USA, May 17-22, (1998).

7. **Grek, F.Z., and V.N. Kisel'nikov,** International Chemical Engineering, **4**, 2, pp263-268, (April 1964).

8. **Bendat, B.S., and Piersol, A.G.,** Random Data: Analysis and Measurement Procedures, 2nd Edition, John Wiley and Sons, New York, (1986).

9. **Nedderman, R.M., and C. Laohakul,** Powder Technology, **25**, pp 91-100, (1980).

10. **Takahashi, H., and H. Yanai,** Powder Technology, 7, pp205-214, (1973).

11. **Abriak, N.E., and R. Gourvès,** Powders and Grains 93, Proceedings of the Second International Conference on Micromechanics of Granular Media, Birmingham, UK, 12 to 16 July, pp 449-450 (1993).

12. **Taylor, E.D., and M.L. Hunt,** Powders and Grains 93, Proceedings of the Second International Conference on Micromechanics of Granular Media, Birmingham, UK, 12 to 16 July, pp 275-279 (1993).

13. **Tallon, S., Davies, C.E.,** "Velocity Measurements in Dense Down Flow of Bulk Solids using a Non-Restrictive Acoustic Method", To be presented at the 1st symposium on on-line flow measurement of particulate solids, University of Greenwich, Kent, UK, 14-15 July, (1998).

Experimental Observation of Pressure Waves in Gas-Solids Fluidized Beds

John van der Schaaf, Jaap C. Schouten, Cor M. van den Bleek
J.M. Burgers Centre for Fluid Mechanics, Chemical Reactor Engineering Section,
Delft University of Technology, Julianalaan 136, 2628 BL Delft, The Netherlands

Pressure waves in gas-solids fluidized beds are generally investigated to characterize gas bubble dynamics, which are important for reactor performance. In literature the origin, propagation and attenuation of pressure waves in gas-solids fluidized beds is still subject of discussion. For Geldart group B particles, pressure waves generated by a gas pulse are characterized as compression waves propagating at velocities of 10 m/s and higher, in agreement with results by Bi et al. [2]. The attenuation of upward traveling pressure waves is determined by the distance of the origin of the wave to the bed surface. Downward traveling pressure waves are not attenuated. In freely bubbling beds a major source of fast pressure waves is bubble coalescence, which generates pressure waves traveling upwards and downwards. The average pressure wave propagation velocity in circulating fluidized beds is adequately predicted by the pseudo-homogeneous flow model for voidages above 0.6.

Research on the origin, propagation and attenuation of pressure waves or pressure fluctuations in gas-solids fluidized beds is generally done to obtain a model or general mechanism describing these fluctuations and their relation to fluidized bed stability and gas bubble dynamics, see e.g. Yates and Simons [1]. The gas bubble dynamics are important for reactor performance. In literature no unambiguous model for the propagation and attenuation of pressure fluctuations in gas-solids fluidized beds is given; most recent articles published on this subject are by Bi et al. [2] and Musmarra et al. [3], and are contradictory. Where Bi et al. support the idea of (pseudo-homogeneous) compressible wave theory, Musmarra et al. subscribe to an elastic wave representation through particle-particle interactions. An adequate model for calculating the attenuation of pressure waves is not given. To elucidate the origin of pressure waves and the mechanism of propagation and attenuation, time series of pressure fluctuations were recorded at different column heights in three types of experiments. Firstly, experiments were performed in a 0.10 m ID fluidized bed of polystyrene particles and of sand particles in which gas pulses were given either under the distributor plate or in the fluidized bed. Secondly, pressure time series were recorded in a 0.40 m ID freely bubbling fluidized bed and thirdly, pressure time series were recorded in a circulating fluidized bed of 1.20 x 0.80 m^2 cross section.

GAS PULSE EXPERIMENTS

The gas pulse experiments were performed in a 0.10 m ID column with polystyrene particles (d_p = 630 µm, ρ_p = 1100 kg/m^3, u_{mf} = 0.14 m/s) and sand particles (d_p = 390 µm, ρ_p = 2650 kg/m^3, u_{mf} = 0.14 m/s) at settled bed heights of 0.390 m. The bed was operated at minimum fluidization velocity when gas pulses of 1.1 10^{-3} m^3/s were either injected in the wind box or at 0.230 m from the distributor plate in the column center. The gas pulses were generated by opening a magnetic valve for a period of 0.010, 0.025, 0.050 or 4 s. Only one bubble was formed at the distributor plate except for a pulse of 4 s long, in which case the fluidized bed became freely bubbling during the pulse time. In each experiment, 40 pulses were recorded. Between each pulse an interval of 4 s was taken to damp out all fluctuations and to let the fluidized bed return to its pre-pulse state. For the 4 s pulse time an interval of 8 s was used. Absolute pressure fluctuations were measured at 0.036, 0.078, 0.230, 0.330 and 0.380 m above the distributor plate when the gas

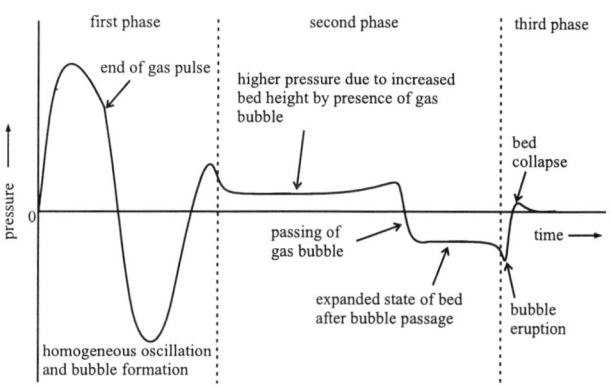

Figure 1 Schematic representation of fluidized bed response to a gas pulse.

pulse was given in the wind box. When the gas pulse was given at 0.230 m above the distributor plate, the absolute pressure fluctuation in the wind box was measured. The sample frequency was 1600 Hz for all experiments and a filter frequency of 800 Hz was used. Typically, the fluidized bed response to a gas pulse can be divided into three phases, schematically presented in Figure 1. First, the bed responds to a gas pulse giving two maxima in pressure, after which the bed gets into an excited state or second phase in which a gas bubble rises to the fluidized bed surface. The bed slowly returns to the pre-pulse state after the bubble has left the bed, which is characterized as the third phase. Information on pressure wave propagation and attenuation is obtained from the first phase.

For the polystyrene particles, the smooth oscillation of pressure in the first phase of approximately 0.20 s is shown in Figure 2 for all heights. Clearly, the maximum

value decreases with distance from the injection position. Attenuation models from Epstein and Carhart [4] or Gregor and Rumpf [5] overestimate the attenuation by orders of magnitude compared to the experimental values; another mechanism is responsible for the pressure wave attenuation. To illustrate the mechanism, the average maximum pressure value in the first phase is plotted versus distance of measurement position to bed surface, $h_{bed} - h_{probe}$, in Figure 3. This results in a straight line with a slope b of $2.45 \cdot 10^3$ Pa/m. The slope b is a measure for the increase in pressure and can be derived from the equation for pressure drop over a packed bed proposed by Ergun [6]:

$$b = \Delta \left\{ \left(150\mu \frac{(1-\epsilon)^2}{\epsilon^3 d_p^2} + 1.75 \rho_f \frac{(1-\epsilon)}{\epsilon^3 d_p} |u_g - u_p| \right) (u_g - u_p) \right\} \quad (1)$$

If the voidage and the particle velocity are taken constant and only a change in gas velocity is assumed, a b value of $3.41 \cdot 10^3$ Pa/m is calculated with Equation (1), which is of the right order of magnitude. The b value is overestimated because changes in particle velocity and voidage have not been taken into account. It can be shown that the pressure dependence on $h_{bed} - h_{probe}$ also holds for values other than the maximum values; the measured local pressure is a direct reflection of the difference in particle and gas velocities and the pressure wave attenuation is linearly dependent on distance to bed surface: $\Delta P(h,t) = b(t) \cdot (h_{bed} - h)$. Consequently, measuring the differential pressure along the height of a fluidized bed will decrease but not filter out fast-propagating pressure fluctuations by gas flow fluctuations. Therefore, differential pressure fluctuations do not characterize gas bubble size or frequency only. The average propagation velocity determined from peak shifts was 55 m/s, which

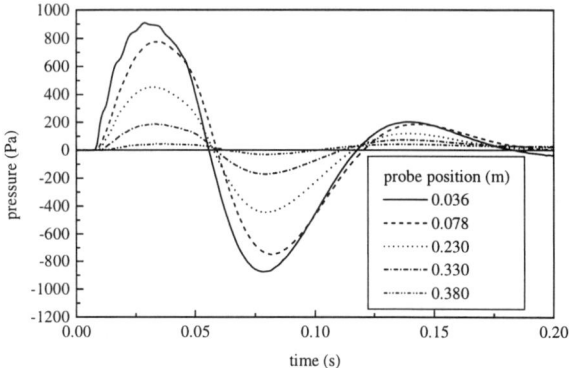

Figure 2 First phase of fluidized bed response for all measurement positions.

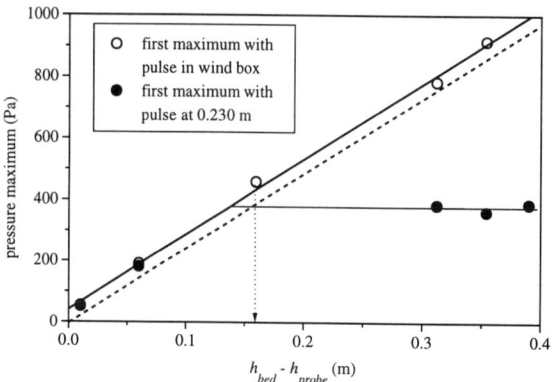

Figure 3 Pressure maximum of fluidized bed response versus distance to surface.

is in the same order of propagation velocities reported by Bi et al. [2] and Musmarra et al. [3]. Because the pressure fluctuations in the first phase are primarily the result of changes in gas velocity, it can be concluded that the propagation mechanism takes place in the gas phase, and the observed pressure wave is thus a compression wave.

Downward propagating waves were investigated with gas pulse injection at 0.230 m from the distributor plate. The values of the first maxima are also plotted in Figure 3. The pressure at all positions below the injection point increase with the same magnitude as the pressure increase at the injection point; a compression wave travels downwards from the injection point on. The amplitude of the downward traveling wave is not attenuated. The pressure values above the gas pulse injection point are equal to the values obtained with gas pulsing in the wind box. Pressure wave attenuation is again determined by $\Delta P(h,t) = b(t) \cdot (h_{bed} - h)$.

The gas pulse experiments described above were also done with a minimum fluidized bed of sand particles to investigate the effect of particle properties on wave propagation velocity and attenuation. The first maximum of pressure in time is now located at 0.036 s with a value of $1.5 \cdot 10^3$ Pa at a height of 0.036 m. The wave propagation velocity in the sand fluidized bed is 36 m/s. Ryzhkov and Tolmachev [7] derived the following equation for the wave velocity when the solids phase and the gas phase are considered separately:

$$u_{wave} = c_0 \left(\frac{\rho_f}{\rho_p (1-\epsilon) \epsilon} \left(\frac{1+\omega^2 \tau_p^2 B}{1+\omega^2 \tau_p^2 B^2} \right) \right)^{\frac{1}{2}}. \quad (2)$$

This model predicts higher wave velocities for higher angular wave frequencies ω. For the limit $\omega \to 0$ or $\tau_p \to 0$, Equation (2) reduces to the equation for the pseudo-homogeneous wave velocity. The wave velocities for the polystyrene and sand fluidized beds of 55 m/s and 36 m/s respectively, are higher than the values predicted by Equation (2) with $\omega = 0$ rad/s (20 m/s and 13 m/s respectively), because of the fast fluctuation in gas velocity. Instead of $\omega = 0$ rad/s (pseudo-homogeneous flow), a typical wave frequency is required as input parameter. As a rough estimation the position of the maximum in the pulse response can be taken as a quarter of the period of the typical frequency. This leads to $\omega = 2\pi/0.021/4 = 75$ rad/s for the polystyrene particles and $\omega = 2\pi/0.036/4 = 44$ rad/s for the sand particles. The calculated wave velocities are then 61 m/s and 26 m/s respectively, which agree better with the experimental values.

FREELY BUBBLING BED EXPERIMENTS

In a 0.384 m ID column, an experiment was performed in a freely bubbling bed of sand particles ($d_p = 390$ μm, $\rho_p = 2650$ kg/m^3, $u_{mf} = 0.14$ m/s) with a settled bed height of 0.34 m and a gas velocity of 0.151 m/s. Absolute pressure fluctuations were measured at the wall at 0.04, 0.09, 0.14, 0.19, 0.24, 0.29, 0.34 and 0.39 m above the distributor plate. A time series of 328 s was recorded with a sample frequency of 1600 Hz and a filter frequency of 30 Hz. In Figure 4, a part of the measured time series is depicted. The pressure fluctuations present in Figure 4 are ascribed to physical phenomena generating pressure fluctuations in fluidized beds, viz., bubble formation, bubble coalescence, and bubble eruption. All these phenomena give rise to a fluctuation in gas flow, thus causing the pressure to fluctuate.

To compare the results for wave propagation velocities from gas pulse experiments with freely bubbling beds, the time series from freely bubbling beds were analyzed by determining time shifts of both maxima and minima of all disturbances with a computer algorithm. This so-called peak shift algorithm searches for a maximum and a consecutive minimum in the time series at a certain position, e.g., the lowest measurement position. Subsequently, the algorithm searches the nearest maximum and minimum in the time series measured simultaneously at a higher position. The maxima and minima of the two time series are considered unambiguously coupled when they comply to a specific set of rules. The most important criterion is that only one maximum must be present in the time interval between the maximum and the minimum in both time series. A more detailed description of the algorithm is given by van der Schaaf [8]. The reciprocal wave

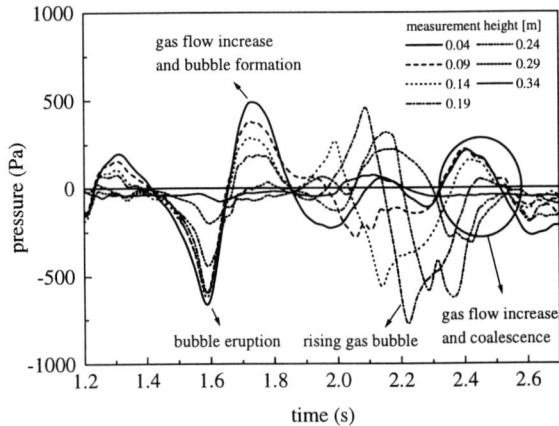

Figure 4 Part of time series measured in 0.385 m ID fluidized bed. Indicated are the different sources for pressure fluctuations.

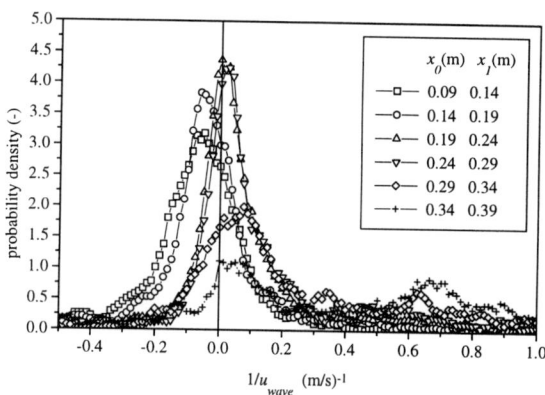

Figure 5 Reciprocal wave velocity distributions in a freely bubbling bed at different measurement positions.

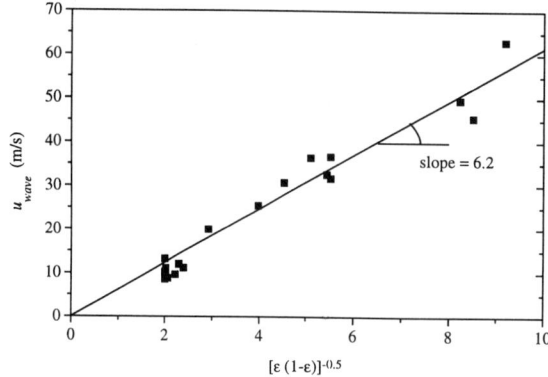

Figure 6 The average wave propagation velocity in a circulating fluidized bed versus $(\epsilon(1-\epsilon))^{-0.5}$.

velocity distributions determined from the time shifts of minima and maxima, are presented in Figure 5. An increase in the distribution can be seen around a value of 0.75 (m/s)$^{-1}$ for the higher probe positions corresponding to rising gas bubbles with a velocity of 1.3 m/s, giving a bubble size of approximately 0.30 m (rise velocity of a single gas bubble $u_b=(gd_b)^{-0.5}$). The average bubble size observed at the bed surface was approximately 0.20 m. Presumably the bubble velocity near the bed surface is higher than predicted for a single rising bubble. The modes of the distributions are near zero. Two explanations can be given for the existence of a reciprocal wave velocity of nearly zero. Firstly, a large bubble can contain two consecutive measurement positions. Because the pressure in the bubble is uniform, disturbances outside the bubble will be detected simultaneously by both probes resulting in a zero reciprocal wave velocity. Secondly, a coalescence of two bubbles generates a pressure wave traveling upwards and downwards. These pressure waves can also be detected simultaneously if the coalescence has taken place between two probes, which again results in a zero reciprocal wave velocity. In Figure 5 it can be seen that below 0.19 m more downward traveling waves are detected than above 0.19 m which implies that bubble coalescence above 0.19 m is a major source of pressure fluctuations. The wave velocity which corresponds to the modes of the distributions determined below 0.19 m and above 0.29 m is approximately 17 m/s, close to the value of 13 m/s calculated with Equation (2) for $\omega = 0$ rad/s.

CIRCULATING FLUIDIZED BED EXPERIMENTS

In a 1.20 x 0.80 m^2 rectangular column of 9 m height, located at Cerchar (Mazingarbe, France), experiments were done with sand particles of 310 µm at a gas velocity of 3.2 m/s. At this velocity, the facility operated at circulating conditions. At the experimental conditions reported here, a dense region with a constant high particle concentration was always present in the riser bottom together with a dilute region with a much lower particle concentration above the dense region. In the dense region large gas bubbles and irregular voids were visually observed in the dense region. The total amount of solids in the column was varied from 568 to 897 kg. The pressure measurement probes were vertically aligned in the center of the 1.20 m wall at heights of 0.28, 0.38, 0.48, 0.58, 1.40 and 1.60 m above the distributor plate. Time series of 15 minutes were recorded with a sample frequency of 800 Hz and a filter frequency of 30 Hz. Next to pressure fluctuations, the average pressure profile over the riser was measured with water manometers, from which the average solids concentration profile was determined. The average wave propagation velocities calculated from the time shifts of maxima and minima is plotted versus $(\epsilon(1-\epsilon))^{-0.5}$ in Figure 6, resulting in a straight line, as expected from Equation (2). The slope is 6.2 ± 0.2 which is reasonably close to the expected value of 7.4 ($\rho_f = 1.08$ kg/m^3, T = 328 K, $\omega = 0$ rad/s). The average wave propagation velocity for low voidages ($\epsilon \approx 0.6$, $(\epsilon(1-\epsilon))^{-0.5} \approx 2$) is overestimated by Equation (2). At these low voidages, gas bubbles or voids with low propagation velocity decrease the average wave velocity. Because the frequencies typically found in fluidized beds are low (<5 Hz), there is no significant difference between the pseudo-homogeneous model and the separate phase model and the pseudo-homogeneous model is adequate in this situation.

CONCLUSIONS

For Geldart group B particles, pressure waves generated by a gas pulse are characterized as compression waves propagating with high velocities, in agreement with results by Bi et al. [2]. The attenuation of upward traveling pressure waves is determined by the distance to the bed surface. Downward traveling pressure waves are not attenuated. Upward traveling pressure waves coincide with bubble formation at the distributor plate; downward traveling waves correspond to bubble eruptions at the bed surface. In freely bubbling beds a major source of fast pressure waves is bubble coalescence, which generates pressure waves traveling upwards and downwards from the origin of the coalescence. The average pressure wave velocity in circulating fluidized beds is adequately described by the pseudo-homogeneous model for voidages above 0.6.

ACKNOWLEDGMENT

The J.M. Burgers Centre for Fluid Mechanics is acknowledged for the financial support (in part) of this project. Cerchar (Centre d'Etudes et de Recherche du Charbon, Mazingarbe, France) is acknowledged for permitting and supporting the measurements in the 0.8x1.20 m² circulating fluidized bed facility.

NOTATION

Roman

b	slope determining pressure wave attenuation	[Pa/m]
B	density ratio (=$\rho_f \epsilon/(\rho_p(1-\epsilon)+\rho_f \epsilon)$)	[-]
c_0	velocity of sound in air (=$(1.4 \cdot R \cdot T/29 \cdot 10^{-3})^{0.5}$ =346 m/s, T = 298 K)	[m/s]
d_p	particle diameter	[μm]
d_b	gas bubble diameter	[μm]
g	gravitational acceleration	[m/s²]
h	distance from distributor plate in fluidized bed	[m]
h_{bed}	bed height at minimum fluidization	[m]
h_{probe}	measurement height	[m]
P	pressure	[Pa]
R	gas constant (= 8.314)	[J/mol/K]
t	time	[s]
T	temperature	[K]
u_b	gas bubble velocity	[m/s]
u_g	superficial gas velocity	[m/s]
u_{mf}	minimum fluidization velocity	[m/s]
u_p	particle velocity in fluidized bed	[m/s]
u_{wave}	pressure wave propagation velocity	[m/s]
x_0, x_1	axial measurement position in fluidized bed	[m]

Greek

Δ	difference with pre-pulse value	[-]
ϵ, ϵ_{mf}	bed voidage, bed voidage at minimum fluidization	[-]
μ	fluid viscosity	[Pa s]
ρ_f	fluid density	[kg/m³]
ρ_p	particle density	[kg/m³]
τ_p	particle relaxation time (= $\rho_p d_p^2/18\mu$)	[s]
ω	angular wave frequency	[rad/s]

LITERATURE CITED

1. J.G. Yates and S.J.R. Simons, "Experimental methods in fluidization research", *Int. J. Multiphase Flow* **20**, pp. 297-330, (1994).

2. H.T. Bi, J.R. Grace and J. Zhu, "Propagation of pressure waves and forced oscillations in gas-solid fluidized beds and their influence on diagnostics of local hydrodynamics", *Powder Technol.* **82**, pp. 239-253, (1995).

3. D. Musmarra, M. Poletto, S. Vaccaro and R. Clift, "Dynamic Waves in Fluidized Beds", *Powder Technol.* **82**, pp. 255-268, (1995).

4. P.S. Epstein and R.R. Carhart, "The Absorption of Sound in Suspensions and Emulsions. I. Water Fog in Air", *J. Acoust. Soc. Am.* **25**, pp. 553-565 (1953).

5. W. Gregor and H. Rumpf, "The attenuation of Sound in Gas-Solid Suspensions", *Powder Technol.* **15**, pp. 43-51 (1976).

6. S. Ergun, "Fluid Flow through Packed Columns", *Chem. Eng. Prog.* **48**, pp. 89-94 (1952).

7. A.F. Ryzhkov and E.M. Tolmachev, "Selection of Optimal Height for Vibrofluidized Bed", *Theoret. Found. Chem. Eng.* **17**, 140-147 (1983).

8. J. van der Schaaf, J.C. Schouten and C.M. van den Bleek, "Origin, Propagation and Attenuation of Pressure Waves in Gas-Solids Fluidized Beds", *Powder Technol.* **95**, pp. 220-233, (1998).

High Order Discretization Methods for the Numerical Simulation of Fluidized Beds

Madhava Syamlal
EG&G, T.S.W.V., Inc., P.O. Box 880, Morgantown, WV 26507

The paper describes the results of a hydrodynamic model of fluidization that uses second-order methods for discretizing the convection terms. The model is used for simulating bubbling fluidized beds and predicts physically realistic "rounded" bubbles. With the same grid resolution, however, the use of a first-order upwind method is the model results in the prediction of bubbles with a "pointed" shape. Therefore numerical diffusion rather than a deficiency in the theory may be the cause of the unphysical "pointed" bubble-shape seen in several published numerical studies.

The paper describes the results of a hydrodynamic model of fluidization that uses second-order methods for discretizing the convection terms. The model is used for simulating bubbling fluidized beds and predicts physically realistic "rounded" bubbles. With the same grid resolution, however, the use of a first-order upwind method in the model results in the prediction of bubbles with a "pointed" shape. Therefore numerical diffusion rather than a deficiency in the theory may be the cause of the unphysical "pointed" bubble-shape seen in several published numerical studies.

Numerical models of fluidized beds based on the multiphase mass and momentum balance equations for gas and solids phases [1] continue to be developed by several groups of researchers around the world. It has been shown that the same set of equations can describe a wide range of fluidization conditions, ranging from bubbling to circulating fluidized beds [1].

The results of bubbling bed simulations, plots of void fraction distribution, show the formation and propagation of high void-fraction regions, called bubbles. Bubble characteristics such as the rise velocity, wake angle, void fraction distribution, and pressure distribution have been compared with experimental data [e.g., 1, 2]. It has been shown that the simulations qualitatively predict experimentally observed solids movement in the bubble wake and *slow* and *fast* bubbles [2].

The predicted bubble shapes observed from contour plots of void fraction differ from experimental observations in one peculiar manner: the nose of the calculated bubble is pointed unlike the rounded shape of experimental bubbles. This problem appears in the simulation results published by various authors using different computational techniques [1, 2, 3, 4, 5]. Another disagreement with experimental observations is that simulations with a continuous jet predict the formation of a permanent "fountain" of solids at the bed surface, much like a spouting bed [6]. To get physically realistic predictions, researchers have attempted to modify the theory. This study shows that these problems are numerical artifacts of using first order accurate discretization schemes and coarse grids and are not due to a fundamental difficulty with the theory.

An implication that the theory can be expected to predict rounded bubbles comes from the numerical simulations based on distinct element models (DEM) [7]. In a DEM model of fluidization developed by Tsuji and coworkers the motion of individual particles rather than the motion of an averaged solids phase is calculated. Gera and Tsuji showed that the DEM model predicted rounded bubble shape, whereas a multiphase model, under identical conditions, predicted a pointed bubble. This comparison implies that the pointed shape is perhaps a numerical artifact, since the solids momentum equation in the multiphase model is a reasonable average of the particle trajectory equations in the DEM model.

The above conjecture was clearly substantiated by Sokolichin et al. in a study of gas-liquid two phase flow [8]. They compared a multiphase flow (Eulerian) model with a Lagrangian particle tracking model. The Lagrangian particle tracking model gave a rounded profile of the gas holdup, where as the Eulerian model yielded a pointed gas holdup profile. They then showed that the multiphase flow model can predict a rounded profile that agrees with the profile calculated with the Lagrangian particle tracking model when the first order upwind (FOU) scheme used in the multiphase flow model is replaced with a second order accurate discretization scheme. The pointed shape is a result of the large numerical viscosity of the FOU scheme, which is of the order of ($U\delta x/2$), where U is the velocity and δx is the grid size. At first it appears counterintuitive that a diffusion effect would make the profile pointed rather than rounded. Sokolichin et al. point out that this so because of the directional dependence of numerical diffusion: the vertical component of velocity is much greater than the radial component, and, therefore, the axial diffusion is much greater than the radial diffusion.

Christie et al. calculated an unpointed, but very unphysical bubble shape, with the FOU scheme. With the use of a second order accurate scheme they could predict rounded bubble shapes [10]. The results of first order upwind scheme and several second order schemes were compared in [11]. The second order methods gave sharper bubble boundaries, but no conclusions were drawn regarding the bubble shape.

This study was motivated by the observation that the shape of the gas hold up profile described by Sokolichin et al.[8] is similar to that of the shape of bubbles in a fluidized bed. Second-order accurate discretization schemes were included in a multiphase flow model of fluidized beds called MFIX [9]. It is shown here that the bubble shape predicted with a second order accurate scheme is rounded. The simulations were conducted for long durations (5 s) and the results did not show the fountain formation at the bed surface. It appears that the fountain formation is caused by coarse grids and low physical viscosity of the solids phase.

DISCRETIZATION METHODS

Fluidized beds are convection-dominated flows, and the convection terms must be discretized accurately. The use FOU scheme for discretizing the convection terms leads to a stable but diffusive set of difference equations. Second and higher order methods, although formally more accurate than FOU, produce spurious oscillations in the solution. To eliminate the spurious oscillations while maintaining higher order accuracy, a *limiter* is applied to the discretization scheme [12]. Second order accurate schemes Superbee, Smart, Muscl, van Leer, and Minmod and the third order accurate method Ultra-quick, were added to MFIX [13]. The *down wind factor* approach discussed in [12] was used for converting the higher order differences into the septa diagonal matrix structure used by MFIX linear equation solver.

TESTING OF DISCRETIZATION METHODS

Figure 1 shows a comparison of a moving plug problem simulated with MFIX to test the various discretization schemes. In this simulation only the solids continuity equation was solved. For clarity, the exact solution and only the solutions obtained with Superbee, Smart and FOU schemes are shown in the figure. At time zero the plug occupies the distance 80-90 cm and moves at a steady velocity of 5 cm/s in the negative x-direction. Figure 1 shows the predictions at 10 s. The FOU and Smart schemes give a smeared, smooth profile. The FOU scheme smears the profile to such an extent that even the peak solids volume fraction is less than 60% of the actual value. The Superbee scheme calculates the exact peak value and nearly reproduces the discontinuity. Ultra-quick results, not shown in the figure, were not distinguishable from the Superbee results. The Minmod and van Leer schemes gave nearly identical results that fell somewhere between FOU and Smart results. A comparison of the normalized CPU times for this simulation is made in Table I.

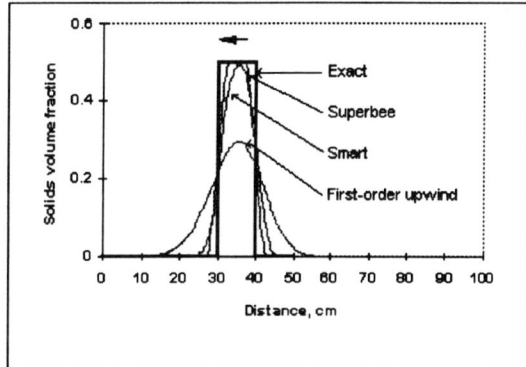

Figure 1. Results of three discretization schemes compared with the exact solution.

Table I
Normalized CPU time requirements for the moving plug problem

FOU	Smart	van Leer	MinMod	Ultra-Quick	Superbee
1	4	5	9	43	81

Table I shows that Smart is the least expensive second order method and Ultra-quick is the cheaper of the two most accurate schemes. When the full set of equations was solved to simulate a bubbling fluidized bed, however, getting a stable solution with Ultra-quick was not possible. Also, the CPU time ratio for Superbee to FOU decreased to a factor of only 1.14. This considerable reduction is perhaps because the ratio of number of computational cells resolving the sharp gradients to the total number of computational cells is smaller in the bubbling bed simulation than in the moving plug simulation. Therefore, in the subsequent simulations Superbee was used as the higher order method of choice, as it gave the most rounded bubble shape.

BUBBLING FLUIDIZED BED WITH A CENTRAL JET

The bubbling fluidized bed experiments of Gidaspow [1, pp. 156-158] with a central jet were simulated. MFIX equations [9] with Gidaspow's drag correlation [1, p. 151] were solved. No symmetry was assumed and a 128x104 cell distribution was used. Figure 2 shows the calculated void fraction (ϵ) distribution in the 500 μm particle bed with a 3.55 m/s

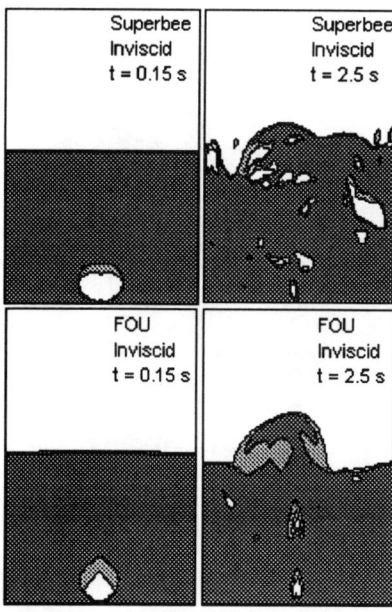

Figure 2. Void fraction distribution in inviscid (solids) simulations.

jet: white - $\epsilon > 0.9$, light gray - $0.7 < \epsilon < 0.9$, dark gray - $\epsilon < 0.7$. In these two simulations the solids viscosity was first set to zero. The first bubble, as it is detaching from the distributor, is shown on the left-hand-side. The nose of the bubble predicted with Superbee is rounded and that of FOU is pointed. With Superbee scheme the bubbles appeared realistic during the first 1 s. After 1 s, the bed became freely bubbling. Some bubbles even moved downwards. A similar loss of bubbling is discussed in [5]. With FOU scheme reasonable behavior lasted up to 2.5 s, and afterwards the bed became freely bubbling (left panels of Fig. 2). The FOU scheme predicts central bubbling for a longer time because of the inherent numerical viscosity.

By turning on the frictional regime stress formulation in MFIX, a stable, centrally bubbling pattern was predicted as seen in experiments (Figure 3). Usually both schemes predict a sharp lower boundary for the bubbles, as shown by the absence of light gray regions near the lower boundary, and a diffused upper boundary. This was also noted in [11]. In Fig 3 the lower boundaries of the bubbles predicted with Superbee, however, have light gray regions because of recent bubble coalescences. The bubbles predicted with Superbee scheme have a much sharper boundary than that predicted with FOU. The rounded bubbles were slightly slower than the pointed bubbles and less frequent (Bubble frequency: FOU - 5 Hz; Superbee - 4.3 Hz).

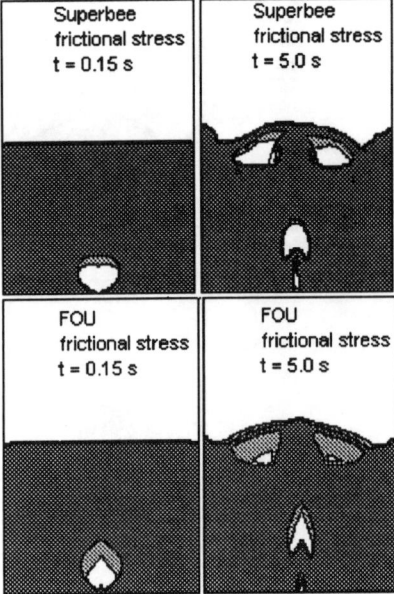

Figure 3. Void fraction distributions with frictional stress terms in solids momentum equation.

Figure 4 shows the simulation results for 800 μm particles and a jet velocity of 5.77 m/s. The comparison between FOU and Superbee schemes show that Superbee scheme predicts a sharper, rounded bubble. The bubble frequencies were comparable: FOU - 7 Hz; Superbee - 6.7 Hz.

Figure 5 shows the void fraction distributions for jet velocities of 5.77 m/s and 9.88 m/s. These simulations also showed that Superbee scheme gives rounded bubble shapes. The bed dynamics become more chaotic at larger jet velocities and bubble shapes deviate from the rounded shape. Also, the bubble being captured in the wake of another bubble becomes elongated. These predictions qualitatively agree with experimental observations. Figure 5 also shows a phenomenon similar to the raining of solids from the bubble roof often seen in experiments [e.g., 1, p.162]. This effect is more clearly seen in animations of the void fraction distributions.

From Figures 2 - 5 also note that a fountain does not form at the bed surface. It seems that the fountain formation is an artifact of the poor grid resolution and the low physical (solids) viscosity in the previous calculations. This problem is more pronounced with the FOU scheme than higher order schemes.

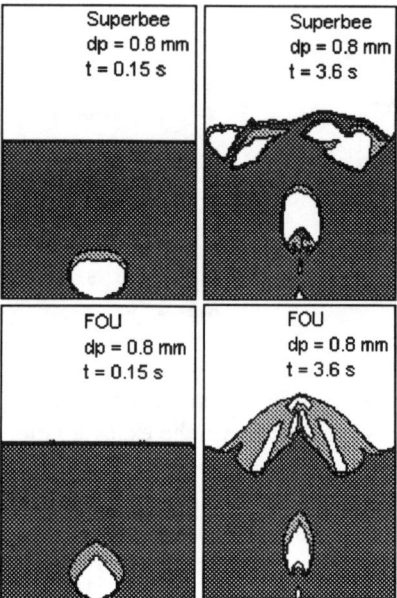

Figure 4. Void fraction distributions for simulations with 800 μm particles.

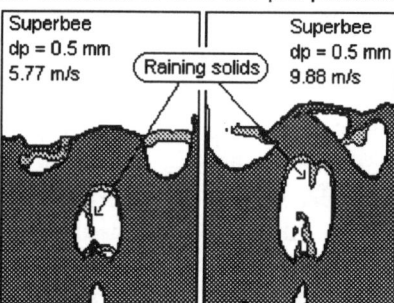

Figure 5. Comparison of void fraction distributions for two jet velocities.

CONCLUSIONS

Five second order accurate discretization methods for convection terms (van Leer, Minmod, Smart, Muscl, and Superbee) were included in a multiphase flow model of fluidization. Several bubbling fluidized bed simulations were conducted without encountering stability problems. The study shows that the unphysical "pointed" shape of bubbles is merely an artifact of the numerical diffusion resulting from the use of first-order upwind scheme. Simulations with the Superbee scheme predict physically realistic rounded bubbles, and appear to be able to mimic details such as the raining of solids from the bubble roof. The simulations were conducted for long durations

(5 s), and there is no evidence of the "fountain" artifact discussed in [6].

LITERATURE CITED

1. Gidaspow, D., Multiphase Flow and Fluidization, Academic Press, New York (1994).

2. Syamlal, M., and T. J. O'Brien, "Computer Simulation of Bubbles in a Fluidized Bed," in Fluidization and Fluid Particle Systems: Fundamentals and Applications, L. S. Fan (Ed), AIChE Symposium Series No. 270, **85**, pp. 22-31 (1989).

3. Kuipers, J. A. M., K. J. Van Duin, F. P. H. Van Beckum, and W.P.M. Van Swaaij, "A Numerical Model of Gas-Fluidized Beds," *Chem. Engng Sci*, **47** (8), pp. 1913-1924 (1992).

4. Boemer, A., H. Qi, U. Renz, S. Vasquez, and F. Boysan, "Eulerian Computation of Fluidized Bed Hydrodynamics -- A Comparison of Physical Models," in Fluidized Bed Combustion - Volume 2, ASME, pp. 775-787 (1995).

5. Sanyal, J., E. Cesmebasi, "On the Effect of various Momentum Transfer Coefficient Models on Bubble Dynamics in a Rectangular Gas Fluidized Bed," *Chem. Engng Sci*, **49** (23), pp. 3955-3966 (1994).

6. Gamwo, I. K., Y. Soong, D. Gidaspow, and R. W. Lyczkowski, "Three-Dimensional Hydrodynamic Modeling of a Bubbling Fluidized Bed," in Proceedings of the 13th International Conference on Fluidized Bed Combustion, K.J. Heinschel (Ed), Book No. H0937A, ASME, pp. 297-303 (1995).

7. Gera, D., and Y. Tsuji, "Hydrodynamics of fluidization, A Legend from Potential Flow Theory to Discrete Particle Simulations," presented at the FED Summer Meeting, ASME, Vancouver (1997).

8. Sokolichin, A., G. Eigenberger, A. Lapin, and A. Lubbert, "Dynamic Numerical Simulation of Gas-Liquid Two-phase Flows, Euler/Euler versus Euler/Lagrange," *Chem. Engng Sci*, **52** (4), pp. 611-626 (1997).

9. Syamlal, M., W. Rogers, T. J. O'Brien, "MFIX Documentation, Theory Guide," Technical Note, DOE/METC-94/1004, NTIS/DE94000087, NTIS, Springfield, VA (1993).

10. Christie, I., G. H. Ganser, J. W. Wilder, "Numerical Solution of a Two-Dimensional Fluidized Bed Model," submitted for publication.

11. Witt, P. J., and J. H. Perry, "A Study in Multiphase Modelling of Fluidised Beds," in Computational Techniques and Applications: CTAC95, R. L. May and A. K. Easton (Ed.), World Scientific Publishing Co., pp. 787-794 (1996).

12. Leonard, B. P., S. Mokhtari, "Beyond First-Order Upwinding: The Ultra-Sharp Alternative for Non-Oscillatory Steady-State Simulation of Convection," *Int. J. Numer. Methods Eng.*, **30**, pp. 729-766 (1990).

13. Syamlal, M., "MFIX Documentation, Numerical Technique," Technical Note, DOE/MC31346-5824, NTIS/DE98002029, NTIS, Springfield, VA (1998).

A Drift-Flux Model for Flow of Nearly-Buoyant Coarse Granular Solids in Liquids

S. Faderani and U. Tüzün
Department of Chemical and Process Engineering, University of Surrey
Guildford, GU2 5XH, UK.

D.L.O. Smith
Process Engineering Division, Silsoe Research Institute
Wrest Park, Silsoe, Bedford, MK45 4HS, UK.

R.B. Thorpe
Department of Chemical Engineering, University of Cambridge
Pembroke St., Cambridge, CB2 3RA, UK.

A mean-field analysis is proposed using Wallis' Drift Flux Model and the Richardson-Zaki correlation to calculate the variation of the plane-mean flowing voidage with mean slip velocity as a function of the terminal velocity of the single particle. For the case of co-current downward flow, the analysis provides a relationship between the solids terminal velocity, the mixture discharge rate and the liquid fraction in the discharge. The resulting expressions can be used to predict, for a specified liquid fraction in the discharge, the corresponding packed-bed to suspension flow transitions and the accompanying voidage changes within a hopper and stand-pipe system.

OBJECTIVE

The purpose of the work is to describe mathematically the plane-mean flowing voidage of a granular solid-liquid mixture discharging from a conical hopper through a vertical stand-pipe.

EXPERIMENTAL STUDY

Granular solid-liquid mixtures are discharged from a model conical hopper and vertical stand-pipe (34 mm and 54 mm diameter). In a series of batch discharge experiments, measurements are made of i) mixture discharge rate and the liquid fraction in the discharge, and ii) dynamic interstitial pressure profiles along the walls of the hopper and the stand-pipe A specially constructed gamma-ray tomographic scanner is used to produce consecutive tomograms of the horizontal planes at selected heights in the hopper and stand-pipe sections. Experiments are repeated for three solid food analogues and one solid model food, each in a mixture with water. The food analogues are extruded plastic particles chosen to represent a range of particle properties, such as size, sphericity, elastic modulus and surface roughness. The model food is uncooked, soaked peas. This work is described in detail in [1].

THE DRIFT FLUX MODEL

The Drift Flux Model, first proposed by Wallis [2], is essentially a separated two-phase flow model where the attention is focused on the relative motion between the phases rather than the motion of individual phases. Here, this approach is particularly appropriate when relating mean slip velocity profiles to the mean voidage profiles along the height of the conical hopper and the stand-pipe sections. Wallis [2] gives the relative flux between the two phases, J_{LS}, in terms of the local voidage, ε, and the relative velocity, $(u_L - u_S) = u_{LS}$, as:

$$J_{LS} = \varepsilon(1-\varepsilon)(u_L - u_S) = \varepsilon(1-\varepsilon)u_{LS} \qquad (1)$$

Before investigating the relationship between the interstitial voidage and the drift flux between the phases in terms of the Wallis model described above, it is appropriate to construct the flow regime transition curves using the Richardson and Zaki correlation for the hindered settling and the minimum fluidisation conditions respectively.

According to Richardson and Zaki [3], the settling speed, u_c, of equal sized particles in a concentrated suspension is related to the single particle terminal velocity by $u_c = u_t \varepsilon^n$, where the exponent, n, is a function of the particle Reynolds number. The slip velocity, Δu, is then related to the particle terminal velocity, u_t, according to:

$$\Delta u = u_L - u_S = u_t \varepsilon^{(n-1)} \qquad (2)$$

Multiplying Equation (2) throughout by $\varepsilon(1-\varepsilon)$ results in:

$$\Delta u \varepsilon(1-\varepsilon) = \varepsilon(1-\varepsilon)u_t \varepsilon^{(n-1)} \qquad (3)$$

which is equal to the drift flux given in Equation (1) above. A dimensionless drift flux can be defined between the liquid and particles by the use of Equations (1) and (3):

$$\frac{J_{LS}}{u_t} = (1-\varepsilon)\varepsilon^n \qquad (4)$$

In the contexts of hindered settling and of incipient fluidisation, the Richardson and Zaki equation is seen to apply. For the case of the incipient fluidisation of a packed-bed, the equation gives rise to the slip velocity between the two phases which is defined by $u_t \varepsilon^{n-1}$. In this case u_{LS} is positive because $u_L > u_S$, as the bed is allowed to expand due to the percolation of the liquid. Thus, the drift flux, J_{LS}, has a positive value over the range of voidages considered. This is shown by the solid line in Fig. 1. Conversely, hindered settling of particles through the liquid results in negative slip (according to Equation (2)) because particles are moving faster than the liquid. This will result in a transition envelope shown by the dashed line in Fig. 1, which corresponds to the negative values of the drift flux in Equation (4).

To predict the interstitial voidage changes that accompany the packed-bed to suspension flow transitions, an equation is required that describes the operating line for a co-current contactor [2]. All quantities are measured as positive in the direction of gravity, this being the direction of flow of the dispersed phase. Assuming that the particles are of uniform size and are uniformly distributed over any cross section of the column, the liquid velocity, u_L, and solid velocity, u_S, relative to the column, are $Q_L/A\varepsilon$ and $Q_S/A(1-\varepsilon)$, where Q_L and Q_S are the volumetric flow rates of liquids and solids.

According to the Wallis' Drift Flux Model, when solids settle in a fluid, an overall mass balance of the two phases reveal that the downwards volumetric flux of the particles is $u_c(1-\varepsilon)$, which must be equal to the liquid flux upwards. Consequently, the mean upward speed of the liquid as it is displaced is equal to $u_c(1-\varepsilon)/\varepsilon$ because the volume fraction of the liquid is ε. At a given position within the flow bed, the velocity of the particles relative to the liquid is therefore given by:

$$u_{LS} = u_c + \frac{u_c(1-\varepsilon)}{\varepsilon} = \frac{u_c}{\varepsilon} \qquad (5)$$

Substituting for u_c from the Richardson and Zaki equation:

$$u_{LS} = \frac{u_t \varepsilon^n}{\varepsilon} = u_t \varepsilon^{n-1} \qquad (6)$$

This result expresses the interaction between the particles and the liquid, which can be written as:

$$u_{LS} = \frac{Q_L}{A\varepsilon} - \frac{Q_S}{A(1-\varepsilon)} = u_t \varepsilon^{n-1} \quad (7)$$

Following Wallis [2], Equation (7) is multiplied throughout by $\varepsilon(1-\varepsilon)$ giving:

$$J_{LS} = u_{LS}\varepsilon(1-\varepsilon) = u_t(1-\varepsilon)\varepsilon^n$$
$$= \frac{Q_L(1-\varepsilon)}{A} - \frac{Q_S \varepsilon}{A} \quad (8)$$

The quantity J_{LS} is called the characteristics velocity, or drift flux, and has an important physical significance. It can be shown [4] that the drift flux is equal to the volumetric flux of the liquid relative to a plane moving at the volumetric average velocity and is:

$$J_{LS} = \frac{W_{mix}}{A\rho_{mix}}[\alpha(1-\varepsilon) - \varepsilon(1-\alpha)]$$

$$= \frac{W_{mix}(\alpha(1-\varepsilon) - \varepsilon(1-\alpha))}{A(\rho_L \varepsilon + \rho_s(1-\varepsilon))} \quad (9)$$

which is the equation of operating line for the co-current flow. In terms of dimensionless drift flux, J_{LS}/u_t, the equation of the operating line becomes:

$$\frac{J_{LS}}{u_t} = \left(\frac{u_{mix}}{u_t}\right)[\alpha(1-\varepsilon) - \varepsilon(1-\alpha)] \quad (10)$$

where $\alpha = fn(u_{mix}/u_t)_o$, which has a simple linear form. However, the fractional liquid discharge, α, is related to the mixture velocity only at the orifice plane as given by an empirical relationship [1]. Hence, for any given mixture discharge rate, there is a single value of the liquid fraction, α, in the discharge, although both u_{mix} and ε vary along the hopper and the stand-pipe. It is now possible to plot the operating line showing the variation of the drift flux as a function of voidage for different values of the mixture discharge rate, W_{mix}, and the corresponding values of α.

The sign of the drift flux, J_{LS}, in Equation (4) depends on the sign convention adopted in relation to the direction of gravity and whether co-current or counter-current motion between the phases is considered. Taking the gravity direction as positive in co-current downward flow, the drift flux, J_{LS}, will be positive when $u_L > u_S$ (i.e. liquid percolating) and negative when $u_S > u_L$ (i.e. particles settling). It is possible to superimpose the operating line for the co-current contactor on the flow regime transition diagram (Fig. 1), as shown in Fig. 2. Above the upper flow regime transition boundary, a packed-bed flow regime exists; within the envelope the particles are accelerating relative to the liquid in the suspension mode until the lower boundary is reached, which marks the free-fall velocity of the solids. On the horizontal axis, J_{LS} is zero, which represents a non-settling suspension; below the horizontal axis, a settling suspension is maintained.

An operating line is fitted through the experimental data points by linear regression [4] The experimental drift flux values mostly fall within the packed-bed and incipiently fluidised flow regimes when the discharge is 'liquid-rich' ($\alpha > 0.5$), whereas the experimental data fall within the suspension flow regime in the neighbourhood of the hopper orifice, which clearly is responsible for the 'solids-rich' discharge ($\alpha < 0.5$).

The gradient of the operating line is a function of the specific value of the discharged liquid fraction, α. In general, the value of the gradient of the operating line is lower for a granular material with a large terminal velocity, which gives rise to small voidage changes in bulk flow.

If the discharge is 'liquid-rich', the gradient of the operating line is rather steep compared to the 'solid-rich' case. Accordingly, the flow regime transitions are coupled with rather small changes in interstitial voidage when the discharge is 'liquid-rich' and much larger voidage changes occur when particles are allowed to settle giving rise to 'solids-rich' discharge.

COMPARISON WITH THE EXPERIMENTAL RESULTS

The interstitial voidage values obtained from the drift flux analysis described above compare well with the height profiles of the plane mean voidage determined from the tomographic scans [1]. This allows for an independent confirmation of the visual determinations of the regions of packed-bed to suspension transitions with different materials and for different mixture discharge rates. Two features are in total agreement with the tomographic observations of the flow regime transitions:

i) The values of the bed voidage obtained corresponding to the zero-slip condition (i.e. non-settling suspension, where $\alpha = \varepsilon$) compare well with the profiles of the mean slip velocity and interstitial voidage.

ii) The maximum values of voidage, determined by the intersection of the operating lines with the terminal settling boundary, all fall below the theoretical limit of 0.8 proposed by Richardson and Zaki [3].

CONCLUDING REMARKS

Successful application of the Drift Flux Model demonstrates that the solid frictional and elastic effects can be ignored totally when dealing with the suspension flow without detriment to the prediction of flowing voidages and interstitial fluid pressures.

The liquid fraction in the mixture discharge will always be greater than 0.5 as long as a packed-bed flow regime can be maintained and significant settling of solid particles is avoided during transport within the hopper and stand-pipe system. Conversely, a fraction of solids greater than 0.5 in the discharge is possible by allowing some settling during transport by reducing the mixture discharge velocity. Significant variations of the flowing voidage with height result from flow regime transitions within the hopper and stand-pipe system; the effect is more strongly manifest at very low mixture discharge rates and for low particle terminal velocities.

The extend of agreement between the results of the drift flux analysis and the other independent measurements of the flowing voidage, fluidisation voidage and particle terminal velocities is very encouraging; implying that even a very simplistic mean field analysis is able to capture almost all the important features of the observed flow phenomena.

REFERENCES

1. Faderani S., Tüzün U., Smith D.L.O., Thorpe R.B., "Interstitial Voidage Effects Governing the discharge and Transport of Nearly Buoyant Granular Solids in Liquids" (1996), AIChE Symposium Series, 92, pp 47-54

2. Wallis, G.B., *One-dimensional two-phase flow*. McGraw-Hill Press, London (1969).

3. Richardson J.F., Zaki W.N., "Sedimentation and fluidization: Part 1." Transactions of the Institution of Chemical Engineers, **32**, 35 (1954)

4. Faderani S., Tüzün U., Thorpe R.B., Smith D.L.O. "Discharge and transport of nearly-buoyant granular solids in liquids - Part II An investigation of the effects of particle properties on mixture discharge rates and pore pressure profiles" (1997) Chemical Engineering Science (accepted)

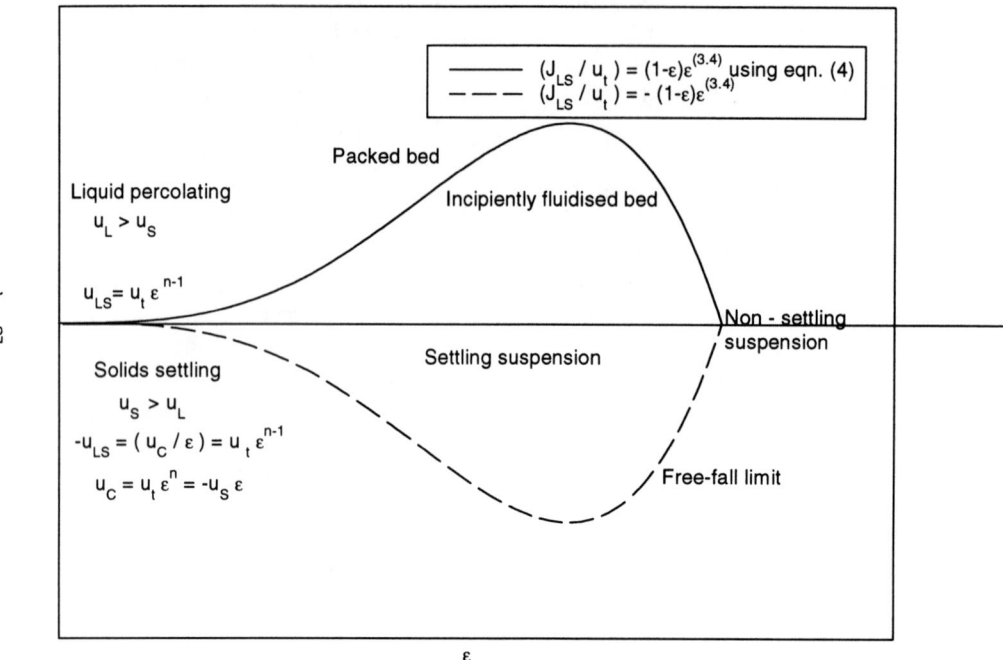

Fig.1 Variation of normalised drift flux, J_{LS}/u_t with voidage, ε

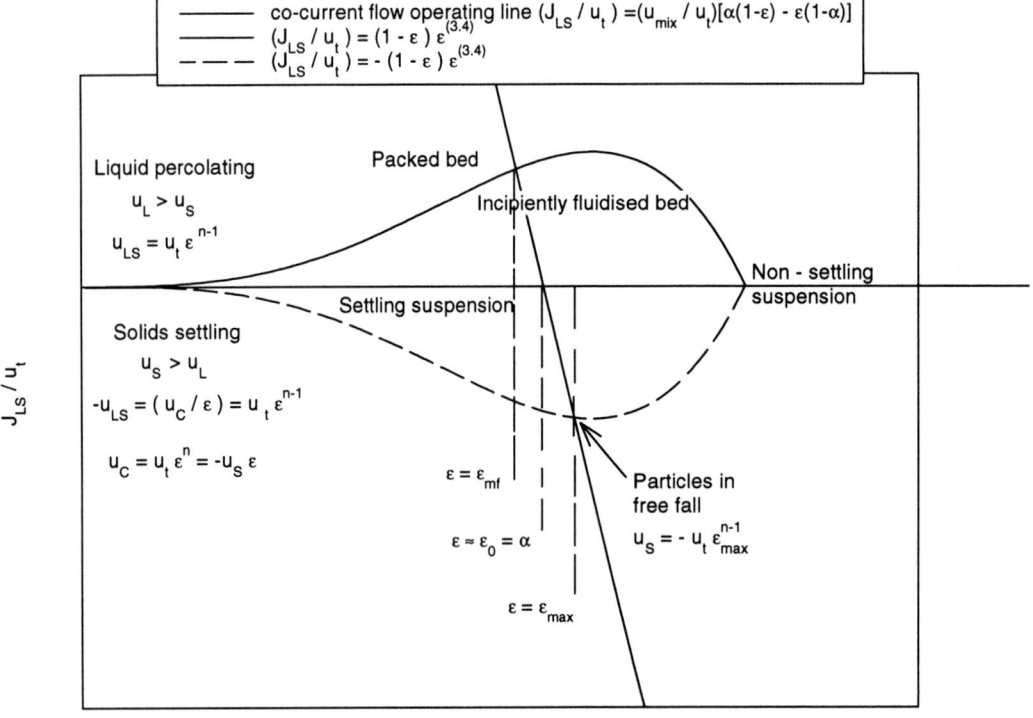

Fig.2 Variation of normalised drift flux, J_{LS}/u_t with voidage, ε, showing the operating line

Modeling Bed-Load Transport by the Master-Equation Approach

L.T. Fan
Department of Chemical Engineering, Kansas State University, Manhattan, KS 66506

Mengtian Bai and Wei-Yin Chen
Department of Chemical Engineering, University of Mississippi, University, MS 38677

The stochastic analysis of sediment and bed-load movements, initiated by Hans A. Einstein, represents a landmark; its significance and utility have been greatly enhanced by the successful development of tracer techniques. This expositional paper reveals that a well-developed stochastic algorithm capable of formulating the master equations of hydrological phenomena yields results sufficiently general as to reduce to those of the special cases previously reported. In the present paper, the location of a particle in the discretized space of interest is considered to be a time-dependent random variable. The transition of the system from one state, i.e., particle location, to another is characterized by a set of intensity functions. The master equations have been established through stochastic population balances of particles in the various states. The analytical solution, derived by the method of eigenvalues and eignevectors, is analogous to the continuous-state dispersion equation of Einstein. Numerical calculations of the evolutions of particle dispersion profiles demonstrate that these two approaches yield comparable results.

By analyzing in detail the locations of particles on the Galton board, Einstein [4] has obtained three equations on the basis of the assumption that both the motion step along the first or spatial coordinate and the rest step along the second or temporal coordinate perpendicular to the first are governed by certain probabilities. The first equation given below represents the conditional probability-density function for the spatial position of a particle after n movements.

$$f(x|n) = k_1 \frac{(k_1 x)^{(n-1)}}{(n-1)!} e^{-k_1 x} \quad (1)$$

In deriving this equation, the step lengths have been chosen to be exponentially distributed with a mean-step length of $1/k_1$. The second equation is the probability-density function of the number of movements, n, at a particular time of t; it is

$$P_n(t) = \frac{(k_2 t)^n}{n!} e^{-k_2 t} \quad (2)$$

In deriving the above expression, the rest periods have been regarded to be exponentially distributed with a mean duration of $1/k_2$. The third equation, the probability-density function of particle position at time t, has been developed to circumvent the difficulty of recording experimentally the number of movements of tracer particles; it is

$$f_x(t) = k_1 e^{-(k_1 x + k_2 t)} \sum_{n=1}^{\infty} \left[\frac{(k_1 x)^{(n-1)}}{(n-1)!} \frac{(k_2 t)^{n-1}}{(n-1)!} \right] \quad (3)$$

In obtaining this equation, the particle is considered to make its first movement at t=0.

Hubbell and Sayre [8] have pointed out that from the standpoint of stochastic mathematics, the gamma distribution of X given by Equation 1 is an expected result since the random variable, X, is the summation of n exponentially distributed random variables, each representing the length of an individual movement. Similarly, the Poisson process of the number of particle movements, i.e., Equation 2, is also expected since the rest periods are exponentially distributed. Hubbell and Sayre have rederived Equation 3 from Equations 1 and 2 through the cumulative distribution function of the conditional probability-density function, Equation 1. Nevertheless, experimental studies by Yang [23] and Grigg [6] have found that a gamma distribution with an additional shape parameter more accurately depicts the observations than an exponential distribution. Subsequently, Yang and Sayre [22] and Shen and Cheong [13] have developed equations for the sand distribution on the basis of this gamma distribution, which are different in form from Equation 3. Todorovic [17] and Shen and Todorovic [16] have reformulated Einstein's homogeneous model by the procedure of Parzen [10]; the resultant set of differential equations is a more general, spatially and temporally nonhomogeneous process. An empirical expression for the time-dependent rate parameter in their model was eventually determined experimentally by Vukmirovic and Wilson [19], and tested against the nonhomogeneous model by Cheong and Shen [2]. Shen and Cheong [13] have also verified mathematically and experimentally that Equation 3 approaches a Gaussian distribution after a sufficiently long time, or $k_2 t \geq 10$. Recently, the applications of Equations 1 through 3 have been extended to processes with the discharge rate-dependent parameters, k_1 and k_2 [15]; to the determination of peak contaminated bed-load profile [14]; and to clogging of the porous column by sediment [21]. Han and He [7] have discussed the relationship between the probabilistic transition rates and fluid mechanics.

In our earlier work [1], the number distribution of particle jumps of Einstein [4], i.e., Equation 2, is rederived by the master-equation approach. In addition, a two-state master equation describing the transitions of particles between the suspended and bed loads has been developed to illustrate the approach's versatility. The present work is concerned with a more general master-equation formulation of the longitudinal dispersion of particles, i.e., Equation 3. Specifically, the aim is to demonstrate that with the same assumption of an exponential distribution for the transition-intensity function as Einstein, the analytical solution of the master equation is analogous to the continuous-state dispersion equation of Einstein. This comparison will indicate the versatility of the former, which has not been fully disclosed in the literature on stochastic hydraulics.

DERIVATION OF MASTER EQUATION

Derivation of the master equation governing the particle dispersions follows what has been established by Oppenheim et al. [9], Gardiner [5], and van Kampen [18]. Let us confine the system of interest to a single particle undergoing successive transitions in the longitudinal direction. The possible range of particle dispersions includes an m-equally-divided domain where the major transitions of particles take place, and the (m+1)th section representing the region beyond the (m)th section where the particle transitions from the first m segments are significantly less important but may not be negligible. The random variable, $N(t)$, denotes the location of the particle. The realization of $N(t)$ varies from 1 to (m+1) which represents the state of the system. The probability that the particle to be in state n at time t is denoted by $P_n(t)$ or $P[N(t)=n]$.

The following assumptions are imposed in deriving the master equation.

1. $N(t)$ is a Markovian random variable, i.e., for any set of successive times $t_1 < t_2 < \cdots < t_q$, we have
$$P[N(t_q)|N(t_1), N(t_2), \cdots, N(t_{q-1})] = P[N(t_q)|N(t_{q-1})]$$
2. The particle movement starts from state 1 at t=0.
3. The number of particle movements depends only on the time interval, i.e., it is temporally homogeneous. For example, $N(\Delta t)$ and $[N(t+\Delta t) - N(t)]$ are identically distributed.
4. The probability of an arbitrary particle to make one movement is proportional to the time interval if the interval is sufficiently small.
5. The particle may move from state i, $1 \leq i \leq m$, to any succeeding state j, $i < j \leq m+1$.

The transition-intensity function, λ_{ij}, is defined on the basis of assumptions 1, 3, 4 and 5 as follows:

Pr [a particle will undergo one transition from state i to state j during the time interval (t, t+Δt)]
= P [N(t+Δt) = j | N(t) = i]
= $\lambda_{ij} \Delta t + o(\Delta t)$ (4)
where

$$\lim_{\Delta t \to 0} \frac{o(\Delta t)}{\Delta t} = 0 \tag{5}$$

The assumptions imposed render it possible to perform a probabilistic population balance around a particular state of the system by taking into account all the mutually exclusive events accompanying the evolution of the system. We consider that the time interval (t, t+Δt) to be sufficiently small so that at most one transition occurs from one lower state to a higher state during this time period. Then, a probability balance yields

$P_n(t+\Delta t)$
= Pr [the state of the system is transformed to n due to a single transition from all possible states below n during the time interval (t, t+Δt)]
+ Pr [the state of the system remains at n during the time interval (t, t+Δt)] (6)

If the two terms on the right-hand side of Equation 6 are designated as Q_1 and Q_2, we have

$$Q_1 = \sum_{k=1}^{n-1} P_k(t) \lambda_{kn} \Delta t + o(\Delta t) \tag{7}$$

$$Q_2 = P_n(t) \left[1 - \sum_{j=n+1}^{m+1} \lambda_{nj} \Delta t \right] + o(\Delta t) \tag{8}$$

Substituting these expressions into Equation 6 yields

$$P_n(t+\Delta t) = \sum_{k=1}^{n-1} P_k(t) \lambda_{kn} \Delta t + P_n(t)\left[1 - \sum_{j=n+1}^{m+1} \lambda_{nj} \Delta t\right]$$
$$+ o(\Delta t) \quad (9)$$

Rearranging the above equation and taking the limit as $\Delta t \to 0$ give rise to the following master equation of the system (see, e.g., van Kampen [18]).

$$\frac{dP_n(t)}{dt} = \sum_{k=1}^{n-1} \lambda_{kn} P_k(t) - P_n(t) \sum_{j=n+1}^{m+1} \lambda_{nj},$$
$$n = 1, 2, \cdots, m+1 \quad (10)$$

Note that the first term on the right-hand side represents probability gain and the second term represents probability loss. By definition, the summation of probabilities of all possible events must be 1, i.e.,

$$P_1(t) + P_2(t) + \cdots + P_m(t) + P_{m+1}(t) = 1 \quad (11)$$

Among the (m+1) equations of Equation 10, therefore, only m equations are independent. Thus, $\mathbf{P}(t)$ defined below constitutes only m independent elements.

$$\mathbf{P}(t) = [P_1(t), P_2(t), \cdots, P_m(t)]^T \quad (12)$$

By substituting this expression into Equation 10, we obtain the first m equations of Equation 10 in a vector form as

$$\frac{d\mathbf{P}}{dt} = \mathbf{W}\mathbf{P} \quad (13)$$

where $\mathbf{W}$ is an (m×m) matrix

$$\mathbf{W} = \begin{bmatrix} -\sum_{j=2}^{m+1}\lambda_{1j} & 0 & 0 & \cdots & 0 \\ \lambda_{12} & -\sum_{j=3}^{m+1}\lambda_{2j} & 0 & \cdots & 0 \\ \lambda_{13} & \lambda_{23} & -\sum_{j=4}^{m+1}\lambda_{3j} & \cdots & 0 \\ \vdots & \vdots & \vdots & \ddots & \vdots \\ \lambda_{1m} & \lambda_{2m} & \lambda_{3m} & \cdots & -\sum_{j=m+1}^{m+1}\lambda_{mj} \end{bmatrix} \quad (14)$$

Assumption 2 suggests that the initial condition for Equation 13 is

$$\mathbf{P}(0) = [1, 0, 0, \cdots, 0]^T \quad (15)$$

Solution of Equation 13 can be found by the method of eigenvalues and eigenvectors. By definition, the eigenvalues of matrix $\mathbf{W}$, ρ_n, can be determined by the equation,

$$|\mathbf{W} - \rho_n \mathbf{I}| = \prod_{n=1}^{m}\left(\rho_n + \sum_{j=n+1}^{m+1} \lambda_{nj}\right) = 0 \quad (16)$$

This expression implies that the eigenvalues of matrix $\mathbf{W}$, ρ_n, are simply the elements on the principal diagonal of $\mathbf{W}$, i.e.,

$$\rho_n = -\sum_{j=n+1}^{m+1} \lambda_{nj}, \qquad n = 1, 2, \cdots, m \quad (17)$$

When the eigenvalues are all distinct, the solution of Equation 13 subject to Equation 15 is (see, e.g., Chiang [3])

$$P_n(t) = \sum_{j=1}^{n} \frac{a_{jn} b_{j1} e^{\rho_j t}}{|\mathbf{B}|}$$
$$= \sum_{j=1}^{n} \frac{a_{jn} b_{j1} e^{-t\sum_{k=j+1}^{m+1} \lambda_{jk}}}{|\mathbf{B}|} \quad (18)$$

where b_{jn} is the n-th element of the eigenvector corresponding to ρ_j; $\mathbf{B}$, the matrix whose columns are the eigenvectors of $\mathbf{W}$; and a_{jn}, the cofactor of the element b_{jn} of matrix $\mathbf{B}$. To compare the results based on the master-equation approach with those of Einstein, it is assumed that the probability of particle transition from one state to any of the higher states depends only on the transition displacement, but independent of the state of the particle. We have, therefore,

$$\lambda_{ij} = \lambda_{i+k, j+k}, \qquad \text{all } j > i \text{ and } (m-i) \geq k \geq 1 \quad (19)$$

and all the eigenvalues of $\mathbf{W}$ are identical, i.e.,

$$\rho_1 = \rho_2 = \cdots = \rho_m \quad (20)$$

For such systems, the solution of Equation 13 subject to Equation 15 has the form (see, e.g., Perko [11]),

$$P(t) = e^{\rho t}\left[I + Nt + \cdots + \frac{N^{m-1}t^{m-1}}{(m-1)!}\right]P(0) \quad (21)$$

In this expression, ρ is the multiple eigenvalue of matrix **W**; **I**, the identity matrix; and **N**, an m × m matrix identical to matrix **W** except that the diagonal elements of the former are all zero.

NUMERICAL SIMULATIONS

As the first step, the relationships have been established between the parameters in Einstein's model, i.e., k_1 and k_2, and those in the master-equation approach. The transition-intensity function, λ_{ij}, defined by Equation 4 delineates the frequency of transition which is expected to decrease as the distance of an individual jump increases. Thus, λ_{ij} is assumed to be in the form of the following exponential function to relate it with k_1 and k_2;

$$\lambda_{ij} = k_4 e^{-k_3[(j-i)h]} \quad (22)$$

where h is the segment length in the discretized spatial domain; thus, (j-i)h represents the displacement, x.

Comparing Einstein's definition of k_1 and the definition of λ_{ij} indicates that

$$k_1 = k_3 \quad (23)$$

The mean frequency of particle jumps, k_2, can be related to the transition-intensity functions, λ_{ij}'s, because it is equal to the reciprocal of the mean rest period as mentioned in connection with Equation 2. Specifically, the probability that a particle will make one jump per unit time is

$$k_2 = \sum_{j=i+1}^{\infty} \lambda_{ij}$$

$$= \sum_{j=i+1}^{\infty} k_4 e^{-k_3[(j-i)h]}$$

$$= k_4 \left[\sum_{j=i}^{\infty} e^{-k_3[(j-i)h]} - 1\right]$$

$$= k_4 \left[\int_0^{\infty} e^{-k_3 h x}dx + \frac{1}{2}(e^0)\right.$$

$$\left. + k_4 \frac{k_3 h}{12}(e^0) - \frac{(k_3 h)^3}{720}(e^0) + \cdots - 1\right]$$

$$= k_4 \left[\frac{1}{k_3 h} - \frac{1}{2} + \frac{k_3 h}{12} - \frac{(k_3 h)^3}{720} + \cdots\right] \quad (24)$$

This expression has been derived by means of the Euler-Maclanrin equation for converting the summation to an integral (Whittaker and Watson [20]). Hence, Equation 24 defines the relationship between k_2 and k_4. The final series expasion in Equation 24 converges faster than the original series, and less terms have to be included in the recovery of k_4.

Once the values of k_3 and k_4 have been determined, the elements in the transition-intensity matrix, **W**, can be calculated from Equation 22. The probability function or simply probability of particle distribution, **P**(t), at any time can then be found by solving Equation 13 numerically, which has been carried out by the 1997- version of ODEPACK (Petzold and Hindmarsh [12]), a software package based on Gear's method for integrating a system of stiff ordinary differential equations. In addition, **P**(t) has been directly computed from the analytical solution, i.e., Equation 21. The results from the former are in good accord with those from the latter. These calculations have been conducted with $k_1 = k_3 = 0.6$ ft^{-1}, $k_2 = 0.042$ h^{-1}, $k_4 = 0.035$ h^{-1} and h = 1ft. The values of k_1 and k_2 have been estimated by Hubbell and Sayre [8] by comparing the flume data with Equation 3. The values of k_3 and k_4 have been estimated by substituting the values of k_1 and k_2 into Equations 23 and 24, respectively. Figure 1 illustrates two discretized distributions at two different times, 77.8 and 125.5h. To cover the entire range of particle distribution in the time domain, 35 spatial sections have been included in the computation.

The results from the master-equation approach discussed above have also been compared with those based on Einstein's model, i.e., Equation 3, as illustrated in Figure 1. For this computation, Equation 3 is integrated in the 35 spatial intervals of interest to convert $f_x(t)$, which is continuous, to the corresponding $P(t)$, which is discrete. The rather small discrepancies between the numerical solution of Equation 13 and the results calculated from Equation 3 are attributable to the fact that the continuous spatial coordinate is discretized in relatively coarse grids in obtaining the former. When the number of spatial sections is doubled, the results from the master-equation approach are indeed in good accord with those from Einstein's model; see Figure 2.

CONCLUSION

Stochastic modeling in terms of macroscopic variables, resorting to both deterministic and probabilistic laws, has been demonstrated to be effective for characterizing the complex dynamics of mesoscopic systems involving particulate matter (see, e.g., van Kampen, [18]). The master-equation approach has been adopted here for modeling the longitudinal dispersion of particles in alluvial streams. The comparison of the results of master equation with those of Einstein demonstrated that, by appropriately defining random variables of interest and the matrix of transition-intensity functions, the master-equation approach can be a highly versatile tool for modeling stochastic phenomena in hydraulics.

NOTATIONS

- a_{jn} Cofactor of element b_{jn}
- b_{jn} Element of matrix **B**
- **B**   Matrix whose columns are eigenvectors of **W**
- f   Probability density function
- h   Segment length for discretization
- **I**   Unity matrix
- k_1, k_2 Parameters in the Einstein model representing the reciprocals of step lengths in the spatial and temporal directions, respectively
- k_3, k_4 Parameters in the exponential form of the transition-intensity function representing the mean frequency and magnitude of particle jumps, respectively
- n   Realization of N, i.e., state of system
- N   Random variable representing particle position
- **N**   Nilpotent of **W**, or an m × m matrix identical to matrix **W** except that the diagonal elements of the former are all zero
- $P(t)$ Probability
- $P_n(t)$ Probability of the particle of interest at state n at time t
- Q_1, Q_2 Probability of two mutually exclusive events described in Equation 6
- T   Superscript representing the transpose of a matrix
- **W**   Transition-intensity matrix
- t   Time
- δ_{ij} Transition-intensity function for movement from state i to state j
- Δ_n n-th eigenvalue of matrix **W**

LITERATURE CITED

1. **Chen, W.Y., L.T. Fan, and M. Bai,** "Applications of the Master-Equation Approach to Sediment Transport Research - Einstein's Model Revisited," Proc. of the Conference on the Management of the Landscape Disturbed by Channel Incision, Oxford, Mississippi, pp.911-916 (1997).

2. **Cheong, H.F., and H.W. Shen,** "Statistical Properties of Sediment Movement," J. Hy. Engineering, **109** (12), 1577-1589 (1983).

3. **Chiang, C.L.,** "An Introduction to Stochastic Processes and Their Applications," Krieger Publishing Co., Malabar, FL (1980).

4. **Einstein, H.A.,** "Bed Load Transport as a Probability Problem,"(1937) in German, Verlag Rascher & Co., Zurich; English translation by W.W. Sayre, in "Sedimentation (Einstein)," H.W. Shen ed., P.O. Box 606, Fort Collins, Colorado, 1972.

5. **Gardiner, C.W.,** "Handbook of Stochastic Methods for Physics, Chemistry, and Natural Sciences," 2nd ed., Springer-Verlag, Berlin, Germany (1985).

6. **Grigg, N.S.,** "Motion of Single Particles in Sand Channels," Ph.D. Dissertation, Civil Engineering, Colorado State University, Fort Collins, CO (1969).

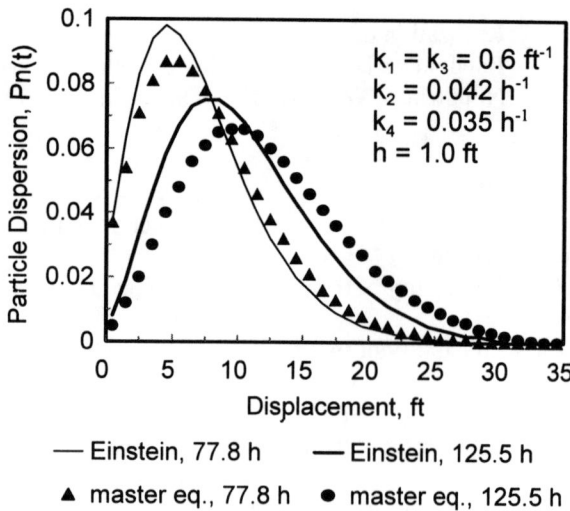

Figure 1. Comparison of the two simulation approaches with 35 spatial sections.

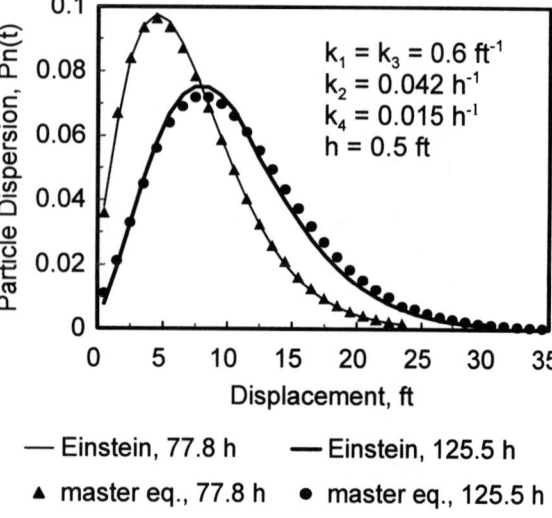

Figure 2. Comparison of the two simulation approaches with 70 spatial sections.

7. **Han, Q., and M. He,** "Stochastic Theory of Sediment Transportation," in Chinese, Publishing House of Science, Beijing, China (1984).

8. **Hubbell, D.W., and W.W. Sayre,** "Sand Transport Studies with Radiaoactive Tracers," J. Hy. Div., ASCE, **90**, HY3, 39-68 (1964).

9. **Oppenheim, I., K.E. Shuler, and G.H. Weiss,** "Stochastic Process in Chemical Physics: The Master Equation," The MIT Press, Cambridge, MA (1977).

10. **Parzen, E.,** "Stochastic Processes," Holden-Day, San Francisco, CA (1962).

11. **Perko, L.,** "Differential Equations and Dynamical Systems," 2nd ed., Springer-Verlag, New York, pp.32-33 (1996).

12. **Petzold, L.R., and A.C. Hindmarsh,** "A Systematized Collection of ODE Solvers," Report of Lawrence Livermore National Laboratory to the U.S. Department of Energy under contract W-7405-Eng-48, also available on web at http://www.netlib.org/odepack/doc (1987).

13. **Shen, H.W., and H.F. Cheong,** "Dispersion of Contaminated Sediment Bed Load," J. Hy. Div., ASCE, **99**, HY11, 1947-1965 (1973).

14. **Shen, H.W., and J.T. Shiau,** "Determination of Peak Contaminated Bed Load Profile," in Stochastic Hydraulics '96, A.A. Balkema, ed., Rotterdam, Netherlands, pp.47-51 (1996).

15. **Shen, H.W., and G.Q. Tabios,** "Movement of Sediment Using a stochastic Model with Parameters Varying as a Function of Discharge," in Stochastic Hydraulics '96, A.A. Balkema, ed., Rotterdam, Netherlands, pp.233-239 (1996).

16. **Shen, H.W., and P. Todorovic,** "A General Stochastic Sediment Transport Model," in Stochastic Hydraulics '71, Chiu, C.L., ed., University of Pittsburgh, pp.489-503 (1971).

17. **Todorovic, P.,** "On Some Problems Involving Random Number of Random Variables," The Annals of Mathematical Statistics, **41**, 1059-1063 (1970).

18. **van Kampen, N.G.,** "Stochastic Process in Physics and Chemistry," 2nd ed., Elsevier, Amsterdam, Netherlands (1992).

19. **Vukmirovic, V., and G. Wilson,** "Bed Load Movement as a Random Process," in Hydraulic Problems Solved by Stochastic Methods, Proceedings of the 2nd Int'l IAHR Symp. on Stochastic Hydraulics, Lund, Sweden, pp.14-1 - 14-26 (1976).

20. **Whittaker, E.T., and G.N. Watson,** "A Course of Modern Analysis," 4th ed., Cambridge University Press, London, England, pp.127-128 (1927).

21. **Wu, F.C., M.H. Hsu, H.W. Shen, and D. Ma,** "A Stochastic Model for Clogging of Porous Column by Sediment," in Stochastic Hydraulics '96, A.A. Balkema, ed., Rotterdam, Netherlands, pp.441-447 (1996).

22. **Yang, C.T., and W.W. Sayre,** "Stochastic Model for Sand Dispersion," J. Hy. Div., ASCE, **97**, HY2, 265-288 (1971).

23. **Yang, T.,** "Sand Dispersion in a Laboratory Flume," Ph.D. Dissertation, Civil Engineering, Colorado State University, Fort Collins, CO (1968).

Mass Transfer and Flow Regimes in Three-Phase Magnetic Fluidized Beds

Chia-Min Chen and Lii-Ping Leu
Department of Chemical Engineering, National Taiwan University, Taipei 106-17, Taiwan

The mass transfer in gas-liquid-solid magnetic fluidized beds was examined by using the axial dispersion model, and its flow characteristics were studied. There are three flow regimes in gas-liquid-solid MFBs including dispersed bubble, coalesced bubble and slugging flow. This study analyzed the volumetric mass transfer coefficient k_{La} to quantify mass transfer in three-phase MFBs.

The solid phase consisted of nickle powders with mean diameter 115 μm, the liquid and gas phases were water and air, respectively. An axial magnetic field was supplied by an external solenoid with direct current. The volumetric oxygen mass transfer coefficient k_{La} was determined from measurements of the steady state oxygen profile across the reactor.

Experimental results showed that the boundary between dispersed and coalesced bubble flow was independent of gas flow rate, and the regime of dispersed bubble was expaned 70% in MFBs. The mass transfer rate in MFBs was higher compared to ordinary fluidzed beds at the same superficial liquid velocity and bed voidage.

Introduction

Since the pioneering report by Filippov [1], there have been a large number of publications on magnetic fluidized beds(MFBs). To date, the performance of three-phase MFBs has not been well characterized. Hu and Wu [2] examined liquid-solid and gas-liquid-solid MFBs and determined the effect of magnetic field intensity on overall gas holdup and radial variations of gas holdup. Kwauk et al. [3] examined bubble properties in three-phase MFBs and found that bubble size decreases with increasing magnetic field intensity. This paper presents experimental measurements of gas to liquid mass transfer coefficients and axial dispersions in three-phase MFBs.

Experimental

Figure 1 shows a schematic diagram of the experimental apparatus. Water flowed through a plexiglas tube (inside diameter 27mm) loaded with 0.15 kg spherical nickel powders of average diameter 115 μ m (ranged from 105 to 125 μ m). A perforated plate with fractional free area 5.8% was used as a distributor for the bed. An application of electrical current to the copper coil generates a magnetic field in the axial direction of the column. Compressed air was injected through a stainless-steel ring sparger under the bed. A holding tank containing water was continually sparged with nitrogen gas until the dissolved oxygen concentration was 0-10% of the air-saturation value. The steady state method was used to calculate the gas to liquid mass transfer coefficient k_{La}. A mass balance on oxygen in the liquid phase is shown as:

$$\frac{1}{Pe}\frac{d^2C}{dZ^2} - \frac{dC}{dZ} - St \cdot C = -St \cdot C^* \quad (1)$$

with the following boundary conditions:

$$Z = 0, C = C_0 + \frac{1}{Pe}\frac{dC}{dZ}$$
$$Z = L, \frac{dC}{dZ} = 0 \quad (2)$$

the analytical solution to above differential equation is:

$$C = A \cdot e^{r_1 Z} + B \cdot e^{r_2 Z} + C^*$$
$$r_{1,2} = \frac{Pe \pm \sqrt{Pe^2 + 4St \cdot Pe}}{2} \quad (3)$$

where

$$A = \frac{C_0 - C^*}{(1 - \frac{r_1}{Pe} + \frac{r_1}{Pe}e^{(r_1-r_2)L} - \frac{r_1}{r_2}e^{(r_1-r_2)L})}$$

$$B = \frac{C_0 - C^*}{(1 - \frac{r_2}{Pe} + \frac{r_2}{Pe}e^{(r_2-r_1)L} - \frac{r_2}{r_1}e^{(r_2-r_1)L})}$$

$$Pe = \frac{UL}{D_a \varepsilon}, \quad St = \frac{k_L a L}{U} \quad (4)$$

the mass transfer coefficient and axial dispersion coefficient were obtained by parameter fitting of above solution to the experimental dissolved oxygen concentration profile. A nonlinear statistical regression program (Sherrod,1992) is employed for the parameter fitting, which gives a minimum value of the sum of squares of the error.

Results and Discussion

Axial dispersion

The inert tracer method used to solve the axial dispersion coefficients have been given elsewhere (Chen and Leu, [4]). Figure 2 shows the comparison of axial dispersion coefficient obtained from tracer experiment and this work under the effect of magnetic field intensity, axial dispersion coefficient decreased with increasing H for all liquid velocities in both methods. In low liquid velocity it decreased rapidly with the addition of magnetic field due to the change of bubble flow properties, while in high liquid velocity it decreased rapidly only at high magnetic field intensity. Axial dispersion coefficients obtained from this work were found to differ significantly from those obtained by using inert tracer experiment. The difference can be explained the inert tracer method may not be appropriate for reactors involving interphase mass transfer. Hu and Wu reported that the average gas fraction increased slightly with increasing H. But in contradiction to the findings of Thompson and Worden [5], they found the reason might be due to the different properties of bubble in small and large particle bed.

Gas-to-liquid mass transfer

The effect of liquid velocity on mass transfer coefficient for various gas velocities is shown in Figure 3, the mass transfer coefficient increased as both the liquid and gas velocity increased. Figure 4~Figure 6 show the variation of k_{La} with magnetic field intensity, the increase of k_{La} with H was statistically significant. However, different mechanisms may be responsible for the enhancement effects. The enhancement seen in Figure 4 occurred at low liquid velocity resulted from bubble splitting. Whereas the enhancement seen in Figure 6 occurred at high liquid velocity resulted from bubble splitting and gas channeling. The maximum k_{La} occurred at largest bubble breakup without gas channeling. This is due to increased bubble splitting in MFBs leading to decreased bubble size and increased interfacial area. Hu and Wu reported that the average gas fraction increased with increasing H, leading to that the bubble diameter decreased with increasing H. A similar result was obtained from this experiment, the addition of magnetic field resulted in bed contraction and smaller bubbles. Epstein [6] indicated that the gas bubbles tend to disintegrating in large particle bed and coalescing in smaller particle bed.

The flow regime map is shown in Figure 7. The boundary between dispersed bubble and coalesced bubble is independent of gas flow rate. With the addition of magnetic field, the regime of dispersed bubble expanded due to the bed contraction and enhanced bubble splitting. Figure 8 shows k_{La} as a function of bed voidage (void occupied by gas and liquid) at various H, and it increased with increasing H at the same bed voidage.

Conclusion

(1) The axial dispersion model was used to described the mass transfer and dispersion in three-phase MFBs, a nonlinear search program was employed to evaluated both axial dispersion and mass transfer coefficients.
(2) Gas bubble tended to split with the addition of magnetic field, the interfacial area between gas and liquid became larger and changed the regime of MFBs.
(3) The mass transfer rate increased with increasing magnetic field intensity at the same superficial liquid velocity and bed voidage due to the enhanced bubble splitting.

Notation

C : concentration of oxygen (mg/dm³)
C_0: inlet concentration of oxygen (mg/dm³)
C^* : saturation concentration of oxygen (mg/dm³)
Da : axial dispersion coefficient (m²/s)
d_p : particle diameter (m)
H : applied magnetic field (A/m)
k_{La} : gas to liquid mass transfer coefficient (1/s)
L : axial distance (m)
Pe : axial Peclet number (-)
Re : particle Reynolds number (-)
St : Stanton number (-)
U : superficial velocity (m/s)
U_l : superficial liquid velocity (m/s)
U_g : superficial gas velocity (m/s)
Z : axial position (m)
ε : void fraction occupied by liquid (-)

Literature Cited

1. Filippov, M. V., "The Effect of a Magnetic Field on a Ferromagnetic Particle Suspension bed", *Prik. Magnit. Lat. SSR*, 12, pp. 215 (1960).

2. Hu, T. T. and J. Y. Wu, "Study on the Characteristics of a Biological Fluidized Bed in a Magnetic Field", *Chem. Eng. Res. Des.*, 65, pp. 238-242 (1987).

3. Kwauk, M., X. Ma, F. Ouyang, Y, Wu, D. Weng and L. Chang, "Magnetofluidized G/L/S Systems", *Chem. Eng. Sci.*, 47, pp. 3467-3474 (1992).

4. Chen, C. M. and L. P. Leu, " Hydrodynamic and Dispersion Behavior of Liquid-Solid Magneto-Fluidized Beds", 5th Asian Proc. Conference on Fluidized-Bed and Three-Phase Reactors, pp. 268-273, Hsitou, Taiwan (1996).

5. Thompson, V. S. and R. M. Worden, "Phase Holdup, Liquid Dispersion and Gas-to-Liquid Mass Transfer Measurements in a Three-Phase Magnetofluidized Bed", *Chem. Eng. Sci.*, 52, pp. 279-295 (1997).

6. Epstein, N., "Three-Phase Fluidization: Some Knowledge Gaps", *Can. J. Chem. Eng.*, 59, pp. 649-657 (1981).

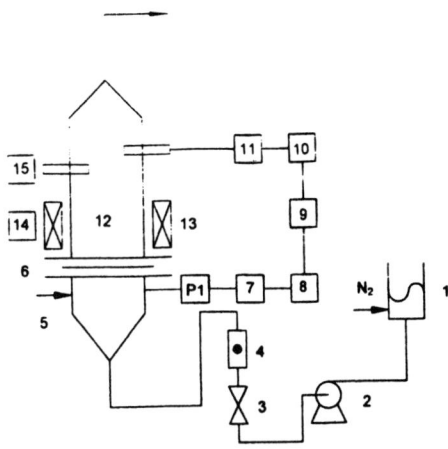

1. stripping tank
2. pump
3. globe valve
4. rotameter
5. gas inlet
6. distributor
7. pressure transducer
8. amplifier
9. data acquisition system
10. computer
11. DO meter
12. fluidized bed
13. magnetic solenoid
14. DC power supply
15. Gauss meter

Fig. 1. Experimental setup

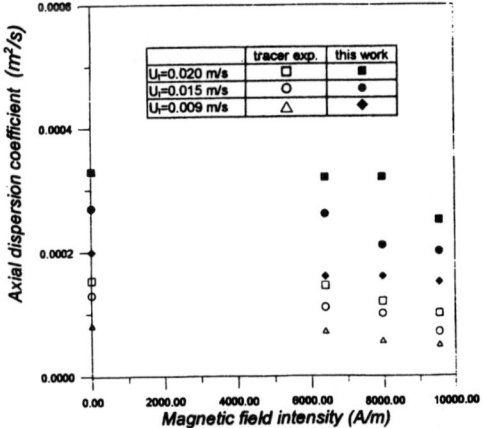

Fig. 2. Comparison of axial dispersion coefficient obtained from profile fitting and tracer experiment (U_g=0.01 m/s).

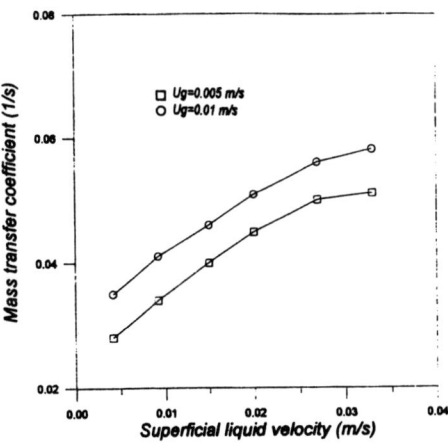

Fig. 3. Mass transfer coefficient as a function of superficial liquid and gas velocity.

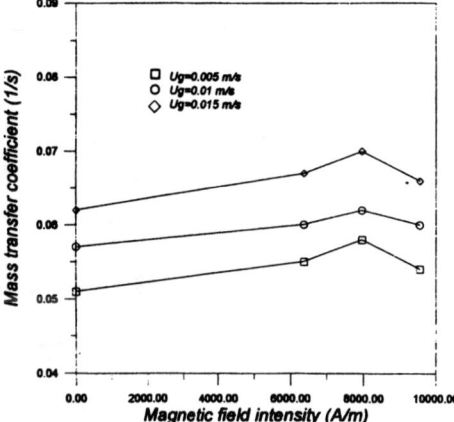

Fig. 4. Effect of magnetic field intensity on mass transfer coefficient (U_l=0.027 m/s).

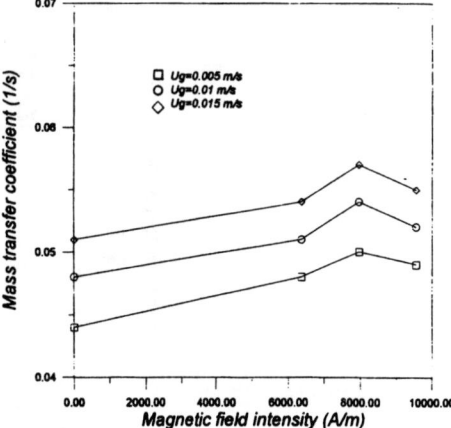

Fig. 5. Effect of magnetic field intensity on mass transfer coefficient (U_l=0.018 m/s).

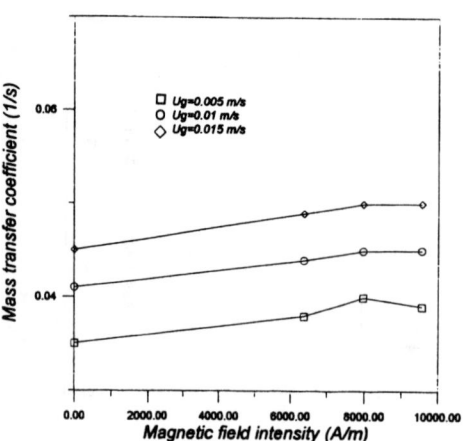

Fig. 6. Effect of magnetic field intensity on mass transfer coefficient (U_l=0.009 m/s).

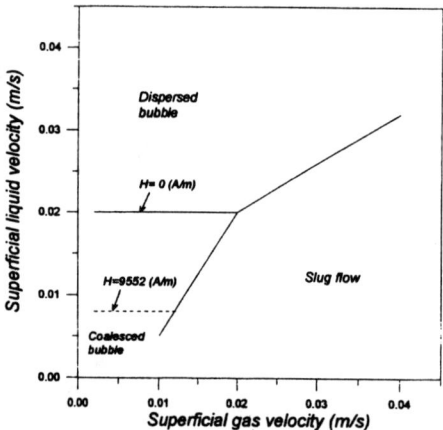

Fig. 7. The moving boundary of flow regime in MFBs.

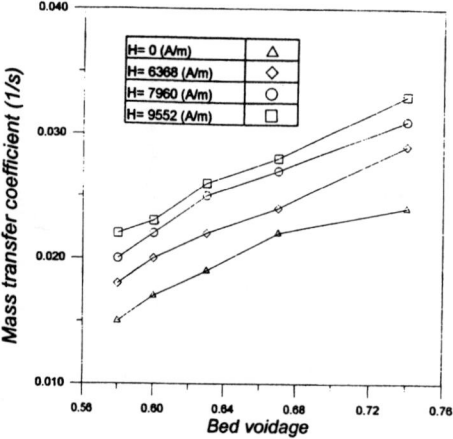

Fig. 8. Mass transfer coefficient as a function of bed voidage under different H.

Characterization of Attrition Properties by a Shear Test

N. Chouteau, P. Guigon, J-F Large
Department of Chemical Engineering, Technical University of Compiégne,
BP 20529, 60205 Compiégne, France

Two attrition tests were developed to isolate the two key mechanisms of attrition: an impact test should be involving the fragmentation of particles, and a shear cell is used to perform their abrasion. This work concerns mainly the second test. A preliminary study allowed first to determine the solid volume to be introduced in the cell. Then, salt particles were tested to study the effect of operating parameters on attrition. This lead to the establishment of an energy criterion adapted to compare product attritabilities. Finally, we observed that the classification of several solids given by the shear test was in agreement with results of an industrial test.

The control of attrition in industrial processes is an important concern which is raising interest because of the new emphasis put on product quality and because of strict environmental restraint concerning pollution by dust.

Two main fundamental mechanisms seem to be involved in particle attrition : fragmentation and abrasion. In order to better identify their importance in the attritability of different powders, two tests have been developed :
- an impact test which has been previously described [1] and is supposed to characterize the attrition mainly due to fragmentation,
- a shear test which is probably more related to abrasion phenomena.

This paper deals with experimental results obtained in a rotating cell by shearing a bed of salt and coal particles and a comparison of these results with the ones obtained in a more conventional industrial test.

EXPERIMENTAL SET-UP

The annular shear cell is based on the same design as the apparatus developed by Bridgwater [2]. The solid is filled in the annular space (1). The annulus is 20 mm wide, having inner and outer diameter of 120 and 160 mm. The lower part (7) is fitted to a base element (6) coupled to a motor which provides rotation in the range 0.03-50 rpm. The upper part (2) presses on the bed of particles and a pneumatic jack allows to apply a vertical force measured by a force sensor in the range 0-400 daN. This part remains stationary by means of a lever arm (3). The vertical aligning of the equipment is performed by air bearing (5).

The slippage of particles on the top and the bottom of the annulus is avoided by two rings (8) whose surface is machined into radial grooves.

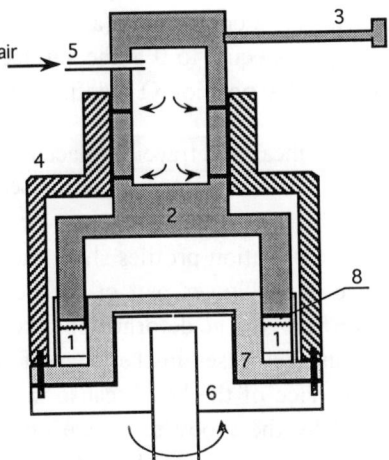

Figure 1 : *Annular shear cell*

DETERMINATION OF THE DEFORMATION ZONE

The preliminary study, described below, has been performed in order to characterize the reference volume of solid to be introduced in the cell in relation with the height of particles in motion and which are, by consequence, submitted to attrition effects. This study is based on tracer experiments carried out with sodium chloride crystals.

A bed of dyed particles is put in the cell, as shown in figure 2, and then is submitted to a shear, under constant vertical force, during 1 or 1/4 revolution. The upper part of the cell is removed and the solid is taken by pumping layers of particles from the rotating lower part. A 1mm high layer is removed at each revolution, and the radial position and dispersion of the tracer can be marked out.

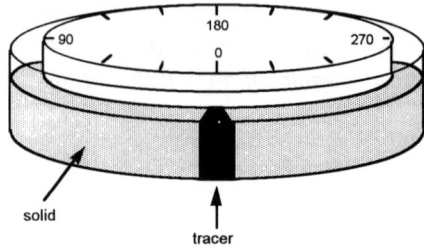

Figure 2 : *Initial position of the tracer column*

The mean position of the tracer (noted Ω) is plotted against its depth in the bed. Figures 3a and 3b show results for two different weights (and heights) of solid. The solid depth is equal to 0 at the surface of the bed, and the mean radial position Ω is defined by :

$$\Omega = \frac{\text{measured tracer displacement}}{\text{total displacement of the cell}}$$

The two deformation profiles show the existence of a dead zone in the lower part of the bed, where the particles don't move. The deformation takes place in the upper part and decreases in the layer of solid located just at the surface of the bed because the particles are well gripped by the grooves. So we observe that the deformation zone does not depend on the height of solid and represents in each case about 5 mm, i.e. 10 particle diameters. The same value has been found for other tested normal pressures (100 and 150 daN).

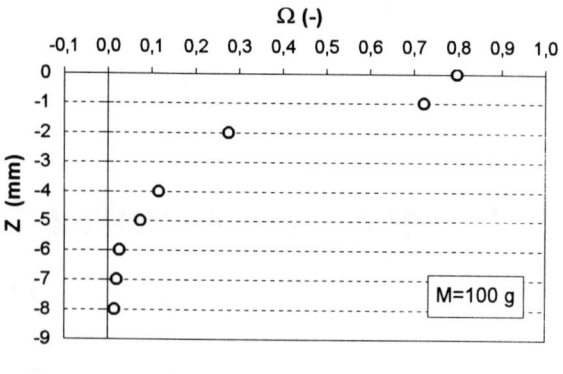

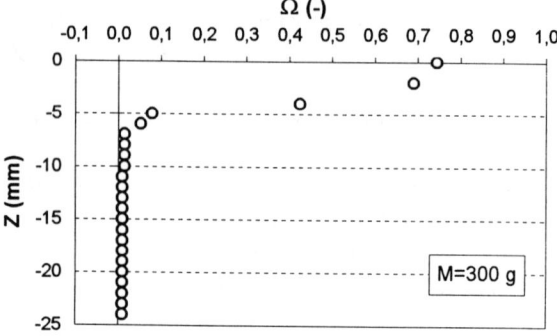

Figures 3a and 3b : *Mean position of the tracer vs its depth for two different weights of solid (NaCl 300-400 µm, F=50 daN, 1/4 revolution)*

EFFECT OF OPERATING PARAMETERS

Effect of Normal Pressure

In figure 4, it is observed that the rate of surface creation ($\Delta S_w/S_w$) is a linear function of the applied normal force. It is measured between two operating limits .
- a lower one which corresponds to the weight of the upper part of the cell (about 4 daN)
- and an upper one which depends on the solid and corresponds to the force at which agglomeration of particles occurs (about 100 daN for NaCl).

Attrition due to shearing mechanisms is represented on figure 5 by particle size distributions at different vertical applied pressures. Compared to the initial distribution, at increasing pressure, a new population of particles appears which corresponds to small fragments of mother particles. In addition to these, fines, less than 1 µm, are identified on SEM photographs but cannot be easily quantified by analysis because of their adherence to the coarser particles. This phenomenon is attributed to the surface abrasion of the particles.

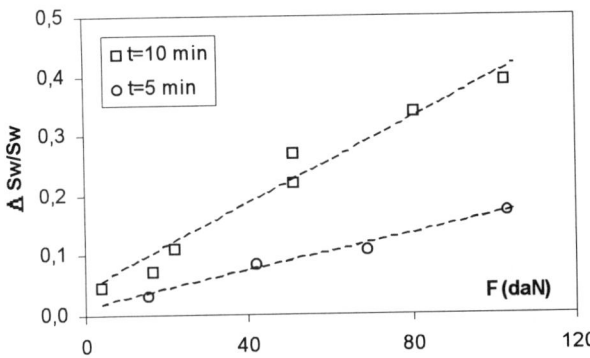

Figure 4 : *Rate of surface creation vs vertical force (NaCl 315-500 µm, ω=1 rev/min)*

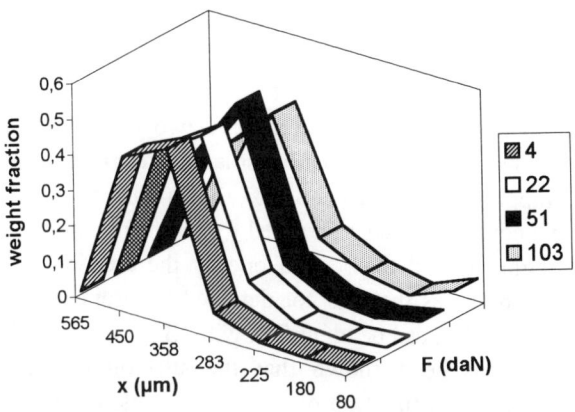

Figure 5 : *Evolution of the particle size distributions (NaCl 315-500 µm, ω=1 rev/min, t=10 min)*

Figure 6 : *SEM photographs of NaCl particles after shearing (1 rev/min, 10 min, 100 daN)*

Effect of Rotation Velocity and Cell Displacement

At first, experiments were carried out at constant cell displacement. In such conditions, the attrition rate does not depend on the rotation speed (figure 7). But when the attrition rate is plotted as a function of the displacement, it varies linearly (figure 8). This means that only the displacement parameter has an effect on particle attrition. The speed and the time of rotation have not.

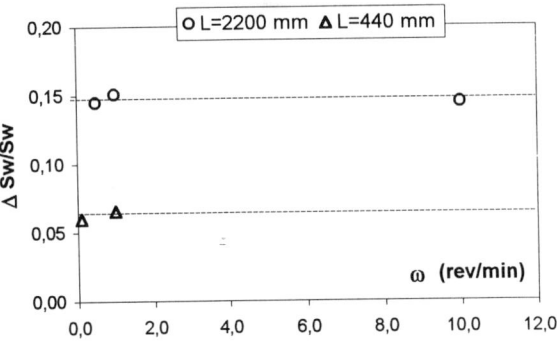

Figure 7 : *Attrition rate vs rotation speed for two displacements (NaCl 315-500 µm)*

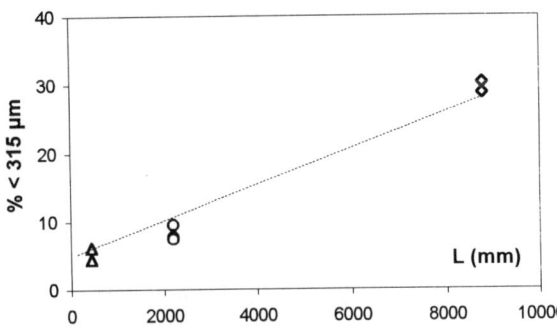

Figure 8 : *Attrition rate vs displacement for different combinations "ω x t" (NaCl 315-500 µm)*

Energy Criterion

The two independent processing parameters which influence attrition are the force and the displacement. In order to compare different products, we have introduced an energy criterion defined by the product of these two parameters. When we plot the created surface vs this criterion (figure 9), all the dots are on the same line. The linearity of the plots allows to compare the behavior of particles of different size, because their slope represents the created surface per energy unit (figure 10). It can be seen that the finest particles are more resistant to wear than the coarsest ones.

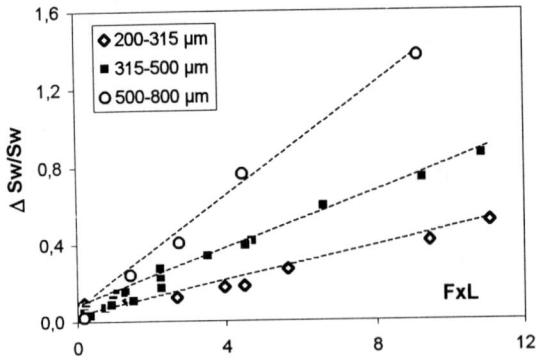

Figure 9 : *Attrition rate vs energy criterion for different size cuts of NaCl*

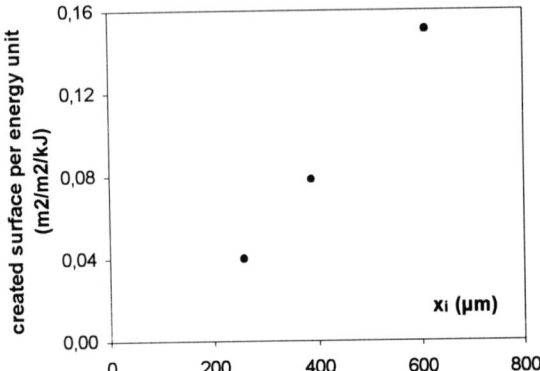

Figure 10 : *Created surface per energy unit vs initial particle size (NaCl)*

COMPARISON WITH AN INDUSTRIAL TEST

The aim of this part is to show the consistency of the shear test results with those of an industrial test. Experiments have been carried out with three types of coal (noted T, G and Y), having different Hardgrove indexes [3]. The Hardgrove test is a standardized procedure of ball milling which leads to the calculation of the Hardgrove Grindability Index (HGI). On the HGI scale, the greater is the HGI value the greater is the grindability.

The results of similar experiments to those described previously are summarized in figure 11. The classification of the solids given by the attrition shear test is in agreement with the Hardgrove data, when the particle size is large enough. Indeed, the particle size has a very important effect on the test sensitivity. The distinction between the three coals is not possible for 400 µm particles, but the classification becomes easy for larger particles (600 µm).

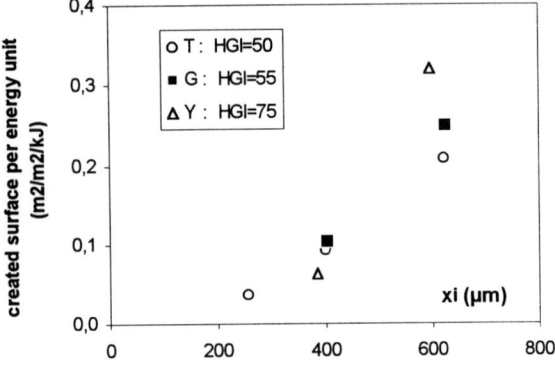

Figure 11 : *Created surface per energy unit vs the initial particle size*

CONCLUSION

The shear test involves two main attrition mechanisms which can't be isolated : **fragmentation** by chipping and **abrasion** of the particles. The first phenomenon is characterized by the formation of small fragments and the second one by the creation of very fine particles that are not detected by ordinary sieving analysis but can be identified with SEM photographs.

It has been shown that the attrition rate can be express as a function of two parameters, the force and the displacement. The linear variation of the particle created surface as a function of a "Force x Displacement " energy criterion is an original result, which has been verified for all the tested solids.

Finally, the shear test is a suitable tool to compare different products and the results obtained in the shear cell are consistent with the ones obtained in other industrial tests.

NOTATION

Z : depth of tracer particles in the bed (mm)
Ω : relative mean radial position or tracer particles (-)
F : vertical force applied on the solid bed (daN)
M : total mass of solid introduced in the cell (g)
ω : angular cell velocity (rev/min)
t : rotation duration (min)
L : cell displacement (mm)

S_w : calculated particle specific surface area (m²/kg)

$$S_w = 6/\rho\, x_{harm}$$

x_{harm} : particles mean harmonic diameter (μm)
ρ : particle density (kg/m³)
$\Delta S_w/S_w$: rate of surface creation (m²/m²)
x : particle size (μm)
x_i : initial mean particle size of the cut (μm)
%<315 μm : creation rate of particles smaller than 315μm (weight %)

LITERATURE CITED

1. **Chouteau, N, et al.**, "Characterization of Attrition Properties by an Impact Testing Technique", *AIChE Symposium series, Progress in Fluidization and Fluid-Particle Systems*, **92**, pp. 106-110 (1996)

2. **Paramanathan, BK, and J. Bridgwater**, "Attrition of Solids-I : Cell Development", *Chemical Engineering Science*, **38**, 2, 197 (1983)

3. **Hardgrove, RM., and P.A. Fullerton**, "Grindability of Coal", *Trans ASME, Fuels and Steam Power*, **54**, 37 (1932)

Index

A
Air injection, secondary 37
Attrition properties 75

B
Bed-load transport 63

C
Circulating beds 37
Closed loop circulating fluidized bed 37
Coarse granular solids 58
Coefficients, lateral dispersion of 20
Computer simulation study 15

D
Dense gas-fluidized beds 15
Dilute region of fluidized beds 20
Discretization methods for simulation 53
Drift-flux model for flow of liquids 58

F
FCC particles, motion of 31
Flow regimes and mass transfer 70
Fluidization
    multi-phase nature of 1
Fluidized beds
    circulating 31, 37
    dense gas 15
    dilute region of 20
    gas 15
    gas-solids 48
    magnetic 70
    numerical simulation of 53
    three-phase magnetic 70

G
Gas-fluidized beds 15
Gas-solids fluidized beds 48
Granular
    dynamics, particle size distribution 15
    flows, slow shearing 25
    solids in liquids 58

H
High-speed imaging 31
Hydrodynamics. 37

L
Lateral dispersion coefficients. 20
Long-range connectivity 25

M
Magnetic fluidized beds 70
Mass transfer and flow regimes 70
Master-equation approach 63
Mechanism of solid flow 37
Modeling bed-lead transport 63
Multi-phase nature of fluidization 1

N
Nearly-buoyant coarse granular solids 58

P
Particles and swarms, FCC 31
Particle size distribution, influence of 15
Pressure waves, observation of 48

S
Secondary air injection 37
Shear test, characterization by a 70
Slow-shearing granular flows 25
Solid(s)
    coarse granular 58
    flow rate 37
    velocity measurement 42

T
Three-phase magnetic fluidized beds 70
Transport, bed-load 63

V
Velocity measurement, solids 42